Theory of Distributions: a non-technical introduction

Theory of Distributions:

a non-technical introduction

IAN RICHARDS

*Professor, School of Mathematics, University of Minnesota,
Minneapolis, Minnesota*

HEEKYUNG YOUN

*Assistant Professor, Department of Mathematics,
College of St Thomas, St Paul, Minnesota*

CAMBRIDGE
UNIVERSITY PRESS

CAMBRIDGE UNIVERSITY PRESS
Cambridge, New York, Melbourne, Madrid, Cape Town, Singapore, São Paulo

Cambridge University Press
The Edinburgh Building, Cambridge CB2 8RU, UK

Published in the United States of America by Cambridge University Press, New York

www.cambridge.org
Information on this title: www.cambridge.org/9780521371490

First published 1990
First paperback edition 1995
Re-issued in this digitally printed version 2007

A catalogue record for this publication is available from the British Library

ISBN 978-0-521-37149-0 hardback
ISBN 978-0-521-55890-7 paperback

CONTENTS

PREFACE

Distributions are sometimes called 'generalized functions', and that is essentially what they are. They correspond to situations presented to us by physical experience which are not adequately covered by the traditional $y = f(x)$ notion of a function. An example is the well-known Dirac Delta Function, which is in fact not a function in the standard sense. The Dirac 'function' corresponds to a unit impulse imparted to a system over what we may idealize as an infinitely short interval of time. Think, for example, of an object being struck by a hammer. While in reality there is some compression of the hammer and of the object, and a small but finite time span during which the interaction occurs, that is not the way we normally see it. To the unaided eye, the whole thing takes place: Bang! – in an instant. This idealization not only corresponds to human intuition, but is very useful in physical applications.

Here an aside. In this discussion, when we use the term 'physical', we really mean 'phenomenological' – i.e. pertaining to the phenomena of nature. Thus, in our usage, the term physical could just as well apply to a problem in mathematical economics as to a problem in mechanics.

This still raises the question: Why create a whole theory to deal with an idea as simple as the Dirac Delta Function? Well, firstly, the idea may not be quite so simple as it looks. More importantly, the idea has important generalizations, each of which could be treated directly on its own merits, but only at the expense of an ever widening loss of clarity and comprehension. When the same idea, in different guises, recurs over and over, there should be an underlying theory which ties the different instances together. The notion of distributions, as introduced and codified by Laurent Schwartz, provides what is today the most widely used of such theories.

This book is addressed to non-specialists. It is intended to be somewhat in the spirit of Lighthill's splendid book [LI]. However, Lighthill uses a non-standard definition of 'distribution'. We believe that the standard approach is, in the long run, more powerful – and of course it has the merit of being standard. On the other hand, we have avoided the use of heavy theories, such as the theory of topological vector spaces. In the first five chapters of the book, we even get by without measure theory (although comments about the measure–theoretic connections occur from time to time).

In the same vein, we have taken great pains to motivate the developments and to make the proofs easy and clear. At times we might be accused of insulting the reader's intelligence. We do not feel that way at all. We assume that our reader is a busy person, who needs to know something about generalized functions, and who wants to learn the basic ideas with a minimum of pain and fuss. To whatever degree this book appears 'elementary', we have succeeded in our intention.

Of course, we do not claim that this book covers everything. There are many deeper results, which the reader can find in more advanced works, that lie outside the scope of ours. For one thing, the theory of topological vector spaces – which we have avoided – eventually comes in at more advanced stages of the theory.

We now give a brief description of the contents.

Chapter 1, as its title suggests, is introductory. It provides the motivation for Chapter 2, Part 1, which introduces the general class of Schwartz distributions. The reader who covers Chapter 2, Part 1, will already have the answer to the question: What is distribution theory about?

Chapter 2, Part 2, deals in a preliminary way with the convolution of distributions. This section is more difficult and could be omitted on first reading. Chapter 3 gives examples, and it could also be omitted on first reading. On the other hand, these examples are so much fun that it would seem rather a shame to ignore them.

Chapter 4 is intended as a preparation for Chapter 5. It gives the required background information on the Fourier transform in its classical setting. Besides merely presenting facts, we have taken pains in this chapter to present the underlying physical motivation.

Chapter 5 is (next to Chapter 2) the most important chapter in the book. It deals with the theory of 'tempered distributions'. Within this theory there is the splendid result that the Fourier transform of a tempered distribution is again a tempered distribution, and that the Fourier Inversion Theorem has universal validity. Anyone who has ever dealt with the Fourier transform in a classical setting – where the Fourier transform of one class of functions is usually a different class of functions, and special definitions are required even to make the theorems make sense – will appreciate the simplicity of the tempered distribution approach. Here again it is in the section on examples that the real fun occurs. One takes the Fourier transform of the function $f(x) = x$ (a function that is not even bounded, much less integrable), and discovers that the Fourier transform is a dipole (a generalized function that is not a function at all). Some of the examples, like the one just cited, are easy; others are harder. One example, starting with the rarified notion of 'generalized function', eventually boils down to a rather tricky computation of gamma function integrals. Finally, we must mention the beautiful result that a tempered distribution is periodic if and only if its Fourier transform is a sequence of delta functions. The coefficients of these delta functions form the Fourier series coefficients for the original distribution. Thus the theory of Fourier series is subsumed – and not by way of analogy but really subsumed – under the theory of the Fourier transform.

Chapter 6 gives the generalizations of distribution theory from the real line $\mathbb{R}^1$ to q-dimensional euclidean space $\mathbb{R}^q$. As noted there, this involves mainly a proliferation of subscripts. All of the basic ideas occur already for $\mathbb{R}^1$ (which is why we wrote most of the book for the one-dimensional case). We mention that a little measure theory is used in Chapter 6, in order to deal with space integrals. However, we have deliberately written the chapter so that an intuitive perception of space integrals should suffice.

Chapter 7 deals with the general problem of multiplication and convolution for distributions. While both multiplication and convolution were defined earlier (in Chapter 2), they were defined only subject to certain side conditions. By the way, these side conditions are absolutely standard – and for good reason: they make the theory easy! The general theory, as laid out in Chapter 7, is more difficult. So far as we know this problem has never been treated in textbook form, although there is a substantial research literature. The approach used here is based on earlier work by one of the authors (Youn) and represents an extension of her Ph.D. thesis.

In conclusion, we repeat a point made earlier. Much of the theory of distributions – the part that most non-specialists need to know – can be done without advanced methods. That is the main theme of this book.

Finally, we wish to thank Gian-Carlo Rota and our editor, David Tranah, for their guidance and patience. Thanks are also due to the staff of Cambridge University Press and the staff of the College of St Thomas for their support. Irene Pizzie of Cambridge University Press brought order to our occasionally chaotic manuscript. Mention must also be made of Susan Moro at St Thomas who did a spendid job of typing the book.

1

Introduction

A 'distribution' is a kind of 'generalized function'. That is, every reasonable function $f(x)$ corresponds to a distribution, but there are distributions which do not correspond to functions. Here we must define what we mean by a 'reasonable' function.

By a reasonable function (or ordinary function or *bona fide* function), we mean a piecewise continuous function $f(x)$ of one real variable x (see Figure 0). This class of functions is sufficient as a starting point. The values of f may be either real or complex numbers, but the variable x is real. In a later chapter we shall extend the theory to several variables.

Now the class of distributions or 'generalized functions' includes many objects which are not functions at all. Why do we study these? The reason is not mere generality. Rather, the theory of distributions has a coherence and power that the classical theory of functions lacks. There are many aspects of this conceptual power which we hope to demonstrate later on, but one

Figure 0. A typical piecewise continuous function $f(x)$. The function is continuous except at the points a_i. For each a_i, the function $f(x)$ has finite left and right hand limits as $x \to a_i$, but these limits may differ (producing the 'jumps'). The value of $f(x)$ at $x = a_i$ itself is immaterial. The partition points a_i are either finite in number, or else approach infinity as $i \to \infty$ and approach minus infinity as $i \to -\infty$.

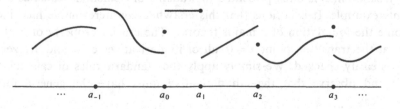

$$\cdots \quad a_{-1} \qquad a_0 \quad a_1 \qquad a_2 \qquad a_3 \quad \cdots$$

of them can be mentioned right away: the operation of taking a derivative applies without restriction to distributions. That is, the derivative of a distribution always exists and is another distribution. By contrast, there are many continuous functions which have no derivative in the classical sense. This defect in the classical approach is eliminated by distribution theory, for, since continuous functions are distributions, the derivative will exist as a distribution.

In this introductory chapter, we intend to lead up to the theory of distributions by a series of intuitive steps based on physical or phenomenological considerations. We interrupt the flow briefly in the next section to give a rigorous definition of the term 'test function'. Since the whole theory is based on test functions, it seemed worthwhile to pin this idea down before proceeding further.

Technical note. For readers familiar with measure theory, we remark that we could replace the class of piecewise continuous functions introduced above (Figure 0) by functions which are locally integrable. One indication of the power of distribution theory is that this extension adds no generality. Every locally integrable function is the derivative of a continuous function!

1 Test functions

This section is slightly technical because of the need to carry out a construction. The reader may prefer to skim it, glance at Figures 1 and 2, and pass on to the next section. As mentioned above, in these early chapters we shall concentrate mainly on functions of one real variable.

A *test function* is a C^∞ function with compact support: 'C^∞' means that the function has continuous derivatives of all orders, and 'compact support' describes functions which vanish outside of some bounded set. The motivations underlying the term 'test function' will be developed as we proceed. But the essential idea is this: curves having certain crude geometrical shapes (like pulses or mesas – see Figures 1 and 2) can be constructed in an infinitely differentiable manner.

We now give the construction. As our point of departure we take

$$h(x) = \begin{cases} e^{-1/x} & \text{for } x > 0, \\ 0 & \text{for } x \leqslant 0. \end{cases}$$

(*The function* $h(x)$ is often presented to students of elementary calculus as a counterexample. It is curious that this erstwhile counterexample has since become the foundation of a major theory.) The crucial property of $h(x)$ is that, at the transition point $x = 0$, all of its derivatives exist and are zero. This is easily checked. We simply apply the standard rules of calculus to $e^{-1/x}$ and observe that the nth derivative must have the general form

$(d/dx)^n(e^{-1/x}) = (a_N x^{-N} + \cdots + a_0) \cdot e^{-1/x}$. No matter what the values of $a_N, \ldots, a_0$ are, the factor $e^{-1/x}$ decreases so rapidly as $x \to 0^+$ that it annihilates all of the other terms.

Now to construct our first basic test function $\varphi(x)$ we set (see Figure 1a):

$$\varphi(x) = h(x)h(1 - x).$$

From this function many others can be constructed. Recall that the *support* of a function f is the closure of the set of x-values for which $f(x) \neq 0$. Thus the function φ in Figure 1a has support on $[0, 1]$. To build a test function with support on any preassigned interval $[a, b]$, we simply set (see Figure 1b)

$$\varphi_{a,b}(x) = \varphi\left(\frac{x - a}{b - a}\right).$$

Figure 1a. A pulse function supported on $[0, 1]$.

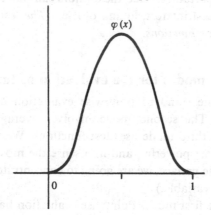

Figure 1b. The same pulse shifted and contracted so that its support is $[a, b]$.

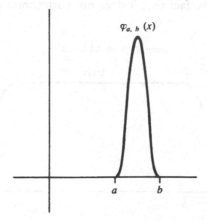

A useful extension of this idea is to integrate the pulse $\varphi_{a,b}$, i.e. to let

$$\psi_{a,b}(x) = \int_{-\infty}^{x} \varphi_{a,b}(u)\,du.$$

Then $\psi_{a,b}$ is a C^∞ function which climbs up from zerto to a positive value and then levels off; the climbing takes place on the interval $[a, b]$. By glueing two such curves together we create a 'mesa function' (Figure 2).

Although we are mainly interested in functions of one variable, we may as well show how to extend these constructions from one to several dimensions. This is easily accomplished by multiplication. For example, to build a pulse on the unit square $\{0 \leqslant x \leqslant 1, 0 \leqslant y \leqslant 1\}$ in $\mathbb{R}^2$, we would take the function $\varphi(x)$ above and then multiply $\varphi(x)\varphi(y)$. Mesas can be extended in a like manner.

Before proceeding further, we should clarify the role that test functions play in the theory. From the viewpoint of applied mathematics, the test functions are not important. In fact, the authors can think of no test function which is of the slightest interest, in and of itself. *The test functions serve as tools in studying other functions.*

2 Three modes for the evaluation of functions

The first mode is the standard pointwise evaluation of $y = f(x)$ at each particular point x. The second mode involves averaging over intervals $\{a \leqslant x \leqslant b\}$, and the third mode uses test functions. We will argue that the third mode is the most powerful, and in a sense the most natural.

(As promised in the preface, we are going to concentrate for the time being on functions of one variable.)

We begin with the first mode. Pointwise evaluation has the advantage of logical simplicity. This may explain why it came first historically, and why it still comes first in everyone's mathematical education. Its defects are a certain rigidity and the fact that it does not correspond to physical reality.

Figure 2. A mesa function.

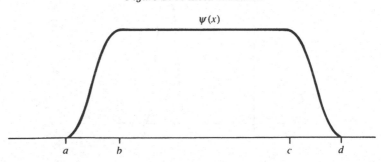

Consider, for example, a chemist studying some property of a substance at temperature T_0. He or she cannot achieve the temperature T_0 exactly, or even achieve a uniform temperature: almost certainly, the temperatures throughout the system will vary over some range $a \leqslant T \leqslant b$. This suggests that the correct mathematical model might involve averaging over $a \leqslant T \leqslant b$ (the second mode above). However, even that is not an accurate description of reality. If a and b are the extreme temperatures, then temperatures close to these extremes will probably occur rarely, and the bulk of the temperatures will be bunched somewhere in the middle. A possible temperature density is indicated by the function $\varphi(T)$ in Figure 3. If the function $\varphi(T)$ is infinitely differentiable, a reasonable hypothesis, then it is a test function.

At this stage we can say a little more about the origin of the term 'test function'. The chemist wants to test the properties of some substance at temperature T_0. So he brings together a batch of the stuff with temperatures distributed in a pulse $\varphi(T)$ near $T = T_0$. The objective is to find some scientific law, and the batch of chemicals (the pulse $\varphi(T)$) tests the law. In a mathematical model, the scientific law will normally be represented by a function, say $f(T)$. Now, just like the chemist, we are interested in f, not φ! We use φ to test f. Let us see how this works out mathematically.

Return briefly to the second mode, averaging. The average value of a function $f(T)$ over an interval $[a, b]$ is $[1/(b-a)] \int_a^b f(T) \, dT$. The division by $(b - a)$ can be carried out arithmetically, so we may as well consider

$$\int_a^b f(T) \, dT.$$

Of course, this corresponds in our model to a uniform temperature distribution over $a \leqslant T \leqslant b$. In the true situation, the temperatures have a

Figure 3. A typical pulse $\varphi(T)$ representing a possible temperature density on the interval $a \leqslant T \leqslant b$. The values a and b are the extreme temperatures, so that $\varphi(T)$ vanishes outside of the interval $[a, b]$.

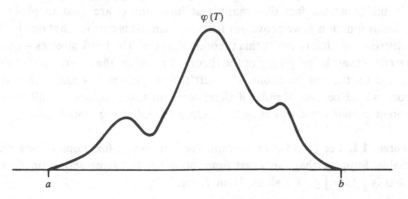

density $\varphi(T)$ supported on $a \leqslant T \leqslant b$. From this we get a 'weighted average',

$$\int_a^b f(T)\varphi(T)\,\mathrm{d}T,$$

in which different temperatures T are weighted according to their frequency $\varphi(T)$ of occurrence. This, of course, is the third mode. It forms the basis for distribution theory.

(From now on we shall drop the second, or averaging, mode. If pushed to its logical conclusion it leads to measure theory. That development is not our objective in this book.)

To summarize our conclusions so far: we are measuring f, and using φ as a tester or probe to do it. Different φ's can probe different temperature ranges. By applying enough probes φ to the same function f, we eventually obtain a knowledge of the structure of f. (Or, as the chemist might say, by enough experiments φ we learn the 'law' described by f.)

There is one technical change we must make in the last formula above, and then we will have our basic definition. Because a and b are the extreme temperature values, there is no harm in extending the integration beyond a or b:

$$\int_a^b f(T)\varphi(T)\,\mathrm{d}T = \int_{-\infty}^{\infty} f(T)\varphi(T)\,\mathrm{d}T,$$

since $\varphi(T) = 0$ for T not in $[a, b]$. It is convenient to use $\int_{-\infty}^{\infty}$ for reasons of homogeneity. But the reader should understand that the integration is really over a finite interval $[a, b]$ determined by the support of φ.

Definition. Let f be a real or complex valued function of one real variable. Suppose that $\int_a^b |f(x)|\,\mathrm{d}x$ exists and is finite for any finite interval $[a, b]$. Then the action of an arbitrary test function φ upon f is defined to be

$$\int_{-\infty}^{\infty} f(x)\varphi(x)\,\mathrm{d}x.$$

(Sometimes for brevity we write $\int f\varphi$ in place of $\int_{-\infty}^{\infty} f(x)\varphi(x)\,\mathrm{d}x$.)

To underline the fact that many test functions φ are used to probe a particular function f, we prove our first theorem. (It is curious that one book on distribution theory omits this theorem entirely. The book stresses – quite correctly – that the simplicity of the theory is based on the severe restrictions imposed on the test functions. It is left for the reader to observe that the theory would be even simpler if there were no test functions at all! Every theorem in that book would be true. Only the following would fail.)

Theorem 1.1. Let f and g be continuous real valued functions of one real variable. Suppose that every test function φ has the same action on f and g, that is $\int f\varphi = \int g\varphi$ for all φ. Then $f = g$.

Proof. The idea of the proof is to construct a narrow 'pulse' $\varphi(x)$ supported on a very short interval $[a, b]$, where $f(x) \neq g(x)$ (see Figure 4). This is what the chemist does when making an accurate experiment, so that the extreme temperatures a and b are very close together. Incidentally, we shall here bid farewell to our hypothetical chemist; the reader has probably grown quite tired of him. Now for the details of the proof.

Suppose $f \neq g$. Then there is some point x_0 where $f(x_0) \neq g(x_0)$, and without loss of generality we can assume that $f(x_0) > g(x_0)$. By continuity, there is some $\varepsilon > 0$ and some $\delta > 0$ such that

$$f(x) \geqslant g(x) + \varepsilon \qquad \text{for } x_0 - \delta \leqslant x \leqslant x_0 + \delta.$$

(Thus for our interval $[a, b]$ we take $[x_0 - \delta, x_0 + \delta]$.) Let $\varphi(x) \geqslant 0$ be a test function (not identically zero) supported on the interval $[x_0 - \delta, x_0 + \delta]$; the existence of such a φ was proved in the previous section. Then,

$$\int_{-\infty}^{\infty} f(x)\varphi(x)\,dx - \int_{-\infty}^{\infty} g(x)\varphi(x)\,dx = \int_{x_0 - \delta}^{x_0 + \delta} [f(x) - g(x)]\varphi(x)\,dx$$

(because the support of φ is on $[x_0 - \delta, x_0 + \delta]$)

$$\geqslant \int_{x_0 - \delta}^{x_0 + \delta} \varepsilon \cdot \varphi(x)\,dx > 0,$$

since $f - g \geqslant \varepsilon$ on this interval, $\varphi \geqslant 0$, and φ does not vanish identically.

Hence $\int f\varphi > \int g\varphi$; that is, φ distinguishes between f and g, as desired.

Remark. Of course, the continuity of f and g is essential in this theorem. If, for example, we changed the definition of g at one point, then the operation $\int g\varphi$ – which depends on integration – would not be altered.

Figure 4. Schematic picture of the proof of Theorem 1.1. We select a small interval $[a, b] = [x_0 - \delta, x_0 + \delta]$, where $f(x) > g(x)$. Then we take a test function $\varphi(x)$ supported on $[a, b]$ and verify that $\int f\varphi > \int g\varphi$.

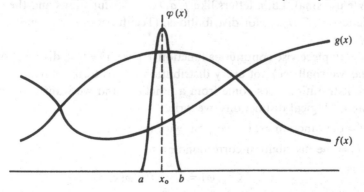

3 Distributions

The time has come to define what we mean by a 'distribution'. The idea is already implicit in the preceding section. We have seen that we can probe a function $f(x)$ (which is the object of study) with test functions $\varphi(x)$, and that each $\varphi(x)$ is mapped into the number $\int f\varphi = \int_{-\infty}^{\infty} f(x)\varphi(x)\,dx$. Thus we decide that:

A distribution is a mapping from test functions to numbers.

The above statement leaves something out. Not all mappings qualify as distributions. There are two side conditions. One of these is slightly technical: it involves the idea of a sequence of test functions $\varphi_1(x)$, $\varphi_2(x)$, $\varphi_3(x)$, ... 'converging to zero'. This is spelled out in Chapter 2. Accepting for now that there is such a notion, we make the following tentative definition.

Definition (provisional). A distribution T is a mapping from test functions to numbers with the following properties. (For a distribution T and a test function φ, we write the action of T on φ as $\langle T, \varphi \rangle$.)

(a) (Linearity.) $\langle T, a\varphi(x) + b\psi(x) \rangle = a \cdot \langle T, \varphi(x) \rangle + b \cdot \langle T, \psi(x) \rangle$ for all test functions φ, ψ and all constants a, b.

(b) (Continuity.) If a sequence of test functions $\varphi_1(x)$, $\varphi_2(x)$, ... 'converges to zero' (definition in Chapter 2), then
$$\langle T, \varphi_n(x) \rangle \to 0.$$

Remarks on notation. The 'inner product' notation $\langle T, \varphi \rangle$ is standard in distribution theory. It reminds us that the transition from functions to distributions is given by an integral $\int f\varphi = \int_{-\infty}^{\infty} f(x)\varphi(x)\,dx$. Incidentally, this is a 'real inner product'; i.e. even if the functions f or φ are complex valued, the integral involves no complex conjugate.

The notation $\langle T, \varphi \rangle$ also brings out the 'duality' which is implicit in our definition: just as different test functions φ can act on a single distribution T, different distributions can act on the same test function. Now we make the further notational conventions:

(1) We use small Latin letters like $f, g, h, \ldots$ for functions and the capital letters $S, T, U, \ldots$ for distributions. (Test functions are denoted by φ, ψ, etc.)

(2) Every piecewise continuous function f gives rise to a distribution, but (as we shall see) not every distribution comes from a function. When a distribution does come from a function, and we want to be careful about logical distinctions, we write:

f = the function evaluated pointwise $(x \mapsto f(x))$;

T_f = the distribution corresponding to f,
$$\langle T_f, \varphi \rangle = \int_{-\infty}^{\infty} f(x)\varphi(x)\,dx.$$

As an example of how distribution theory works, we will now see how to take the 'derivative' of a non-differentiable function. We break the discussion into three steps.

Step 1. As our starting point, we ask what the operation would look like if the function f *were* continuously differentiable. This question has already been answered: the interaction between a function $f'(x)$ and a test function $\varphi(x)$ is given by

$$\int_{-\infty}^{\infty} f'(x)\varphi(x)\,dx.$$

Step 2. The next step forms a model for every construct in the theory of distributions. We try to 'get rid of' the derivative operation on f by 'doing something' to the test function φ. This is achieved using integration by parts. Remembering that $\varphi(x)$ (which has compact support) vanishes as $x \mapsto \pm\infty$, we find that

$$\int_{-\infty}^{\infty} f'(x)\varphi(x)\,dx = -\int_{-\infty}^{\infty} f(x)\varphi'(x)\,dx.$$

On the right hand side we have $f(x)$ (without derivative) – the operation of differentiation has been carried over to φ. The derivative of φ exists since φ is C^{∞}.

Step 3. Now, for the last step we consider a function $f(x)$ which may not be differentiable. For simplicity, we assume that f is at least piecewise continuous (see Figure 0). We have seen that if f *were* differentiable, then

$$\int f'\varphi = -\int f\varphi',$$

and we have noted that the right hand side involves only f and not f'. So we *define* the distribution f' to be the operation which carries every test function φ into the number $-\int_{-\infty}^{\infty} f(x)\varphi'(x)\,dx$.

The same idea extends to arbitrary distributions T. Recall that there we write $\langle T, \varphi \rangle$ in place of $\int T\varphi$. Then we define T' by the formula

$$\langle T', \varphi \rangle = -\langle T, \varphi' \rangle.$$

We will now give our first example of a distribution that is not an ordinary function. This is the famous 'Dirac Delta Function' $\delta(x)$ (pictured crudely in Figure 5b). Strictly speaking, $\delta(x)$ is not a function of x at all.

Remark A. (See, however, the antithesis in Remark B below.) In elementary texts and lectures, the delta function is sometimes treated as though it were an ordinary function, taking a value 'infinity' at $x = 0$. The pulse is assumed to have 'infinite height and width zero, but total area = 1'. This means that we are led to a number system with an '∞' so that $\infty \cdot 0 = 1$. Logically this leads to difficulties. For example, if we suppose that $\infty \cdot 0 = 1$, then there must be a number '$2 \cdot \infty$' so that $(2 \cdot \infty) \cdot 0 = 2$. The trouble is that the laws of algebra fail in this system: $(2 \cdot \infty) \cdot 0 = 2$ but $\infty \cdot (2 \cdot 0) = \infty \cdot 0 = 1$.

The theory of distributions provides an easy way out of this impasse. We start with the discontinuous function $F(x)$ shown in Figure 5a:

$$F(x) = \begin{cases} 1 & \text{for } x \geqslant 0 \\ 0 & \text{for } x < 0. \end{cases}$$

Classically the derivative $F'(x) = 0$, for $x \neq 0$, and $F'(0)$ does not exist. Intuitively, however, we could think of $F'(x)$ as being a unit 'impulse' located at $x = 0$. Now, in distribution theory, this intuitive perception becomes a rigorous fact. Since $F(x)$ (not $F'(x)$) is a *bona fide* function, we have, from our distribution–theoretic definition of F',

$$\langle F', \varphi \rangle = -\langle F, \varphi' \rangle$$

$$= -\int_{-\infty}^{\infty} F(x)\varphi'(x)\,dx$$

$$= -\int_{0}^{\infty} \varphi'(x)\,dx \quad \text{(by definition of } F)$$

$$= -\varphi(x)\big|_{0}^{\infty} \quad \text{(by elementary calculus)}$$

$$= \varphi(0) \quad \text{(since } \varphi(x) \text{ vanishes as } x \to \infty).$$

Figure 5. (a) The 'step' function $F(x)$ whose derivative is $\delta(x)$. (b) A schematic picture of the delta function $\delta(x)$. The pulse is to be viewed as 'very thin and very high', with its total area equal to 1. It is centered above the point $x = 0$.

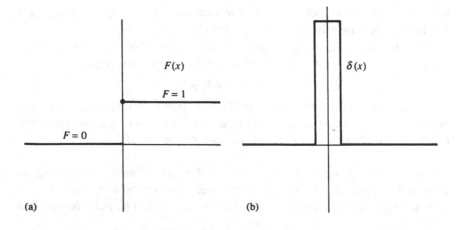

Thus we see that $F'(x)$ (which is usually written $\delta(x)$) is the transformation that maps every test function $\varphi(x)$ into its value at the origin $\varphi(0)$. Intuitively, this is what we would expect the 'unit impulse' at $x = 0$ to do.

Remark B. Our last comment represents an opinion of the authors, which the reader is free to disregard. Students sometimes get tied up in the logical technicalities of this theory and lose the forest for the trees. Logically, a distribution T is a mapping from test functions φ to numbers $\langle T, \varphi \rangle$. However, that is not the right way to think about them. They should be viewed (imprecisely, but suggestively) as 'generalized functions' of x, and written $T(x)$. The test functions φ give a logical underpinning to the theory – but, whenever possible, they should be kept in the background. For example, the (incorrect) picture of the Dirac Delta Function in Figure 5b gives a much better impression of what this construct means than the mathematically rigorous definition $\langle \delta, \varphi \rangle = \varphi(0)$.

(A corresponding anomaly between theory and intuitive perception appears in the definition of real numbers. According to one standard definition, the number π is an equivalence class of Cauchy sequences of rationals. No one thinks of it that way. We think of π as simply 'a number'. Similarly we should think of the delta function $\delta(x)$ as a function of x, even though, strictly speaking, it isn't.)

4 Outline of further developments

As a preliminary to our further developments, we would like to review the steps used to define the derivative of a distribution in the preceding section. However, let us suppose that the operation d/dx is to be replaced by a more general operation, which we call 'Op'. To flesh this out, we list the five main operations that we shall consider. Starting with a distribution T, we want to define:

(1) $(d/dx)T$ (already done above);
(2) $T(ax)$, where $a \neq 0$ is a constant;
(3) $T(x - a)$, where a is a constant;
(4) $g(x)T(x)$, where $g(x)$ is a C^∞ function;
(5) $S(x) * T(x)$, where $*$ denotes convolution, and T is a distribution with compact support.

Since the plan of attack is the same for each case, we use 'Op' to denote any one of the above operations. For example, for (2), $\mathrm{Op}(f)(x) = f(ax)$. In the steps which follow, if we took the first operation d/dx we would virtually be rewriting Steps 1–3 in the previous section. The parallelism is intentional.

Step 1. Start with the classical situation, in which $\mathrm{Op}(f)$ is a *bona fide* function. Add whatever restrictions on $\mathrm{Op}(f)$ we need, e.g. that it is piecewise continuous. Then, by our standard procedure for passing from functions to distributions, Op(f) acts on φ to produce

$$\int \text{Op}(f) \cdot \varphi = \int_{-\infty}^{\infty} \text{Op}(f)(x) \cdot \varphi(x)\, dx.$$

(For example, in the last section Op(f)(x) = $f'(x)$ and the action of f' on φ is given by $\int f'\varphi$.)

Step 2. Try to find another operation Op* which can be applied to test functions and satisfies the identity

$$\int \text{Op}(f) \cdot \varphi = \int f \cdot \text{Op*}(\varphi).$$

For example, in case (1), with Op = d/dx, we have already found that Op* = $-$d/dx; this merely summarizes the integration by parts formula:

$$\int_{-\infty}^{\infty} f'(x)\varphi(x)\, dx = -\int_{-\infty}^{\infty} f(x)\varphi'(x)\, dx.$$

For cases (2), (3) and (4) (omitting convolution for now), we have

$$\begin{aligned}
\text{Op}(f) &= f(ax) & \text{Op*}(\varphi) &= |a|^{-1}\varphi(x/a), \\
\text{Op}(f) &= f(x-a) & \text{Op*}(\varphi) &= \varphi(x+a), \\
\text{Op}(f) &= g(x)f(x) & \text{Op*}(\varphi) &= g(x)\varphi(x).
\end{aligned}$$

To justify this, e.g. for $f(x-a)$, we merely observe that

$$\int_{-\infty}^{\infty} f(x-a)\varphi(x)\, dx = \int_{-\infty}^{\infty} f(x)\varphi(x+a)\, dx.$$

The other two cases are left to the reader.

Step 3. For an arbitrary distribution T, we *define*, replacing $\int T\varphi$ by the distribution notation $\langle T, \varphi \rangle$,

$$\langle \text{Op}(T), \varphi \rangle = \langle T, \text{Op*}(\varphi) \rangle.$$

(Thus, for Op = d/dx, we define $\langle T', \varphi \rangle = -\langle T, \varphi' \rangle$.)

This requires, of course, that the operation Op* maps test functions into test functions. This limitation serves to determine the generality of the various operations we consider. (For example, in (4), multiplication, we need $g(x)$ to be C^{∞}.)

As so often happens, in the logical 'theorem–proof' development things get switched around. We start with Step 3, as a definition. Step 1 becomes another definition. Then we prove a 'consistency theorem' which embodies Step 2. Let us examine more closely the form that a consistency theorem must take. Here it is helpful to be logically precise and distinguish between a (*bona fide*) function f and the distribution T_f which corresponds to it.

(Recall that an ordinary function maps numbers to numbers; a distribution maps test functions to numbers.) The connection, given by Step 1, is:

$$\langle T_f, \varphi \rangle = \int_{-\infty}^{\infty} f(x)\varphi(x)\,dx.$$

Then our consistency theorem would state that, if $Op(f)$ is an ordinary function,

$$Op(T_f) = T_{Op(f)}.$$

In plain English, the distribution definition on the left (where Op operates on the distribution T_f) corresponds to the classical definition on the right (where Op is applied directly to the function f). Of course, we should expect this; otherwise distribution theory would be rather pointless. For example, if we prove by distribution theory that

$$(d/dx)\cos x = -\sin x,$$

we want to know that this formula is really true (i.e. true in the classical sense). This is what the consistency theorem shows: if distribution theory leads to a formula which is classically meaningful, then that formula is also classically true.

On the other hand, part of the charm of this subject is that distribution theory contains formulas which are *not* classically meaningful, e.g.

$$\sum_{n=-\infty}^{\infty} \delta''(x - 2\pi n) = \frac{-1}{\pi}\sum_{n=1}^{\infty} n^2 \cdot \cos nx.$$

Traditionally, this formula makes no sense: the cosine series diverges (its individual terms blow up), and similarly the second derivative of the Delta Function has no classical meaning. However, this is a valid result in distribution theory.

Besides consistency theorems, there is one other type of theorem which we shall consider in the next chapter. Given any two different operations, there may be a formula connecting them. For example, in calculus we have the formula for the derivative of a product: $(fg)' = f'g + fg'$. We would hope that such identities generalize to distribution theory, and usually they do. Theorems to this effect are called 'identity theorems'.

2

The elements of distribution theory

Part 1 Basic definitions and facts

This chapter is self-contained. We refer back to Chapter 1 for motivation, but the reader who prefers a 'definition–theorem–proof' development can simply skim the appropriate passages. We restrict our attention to functions of one real variable. The values of these functions may be real or complex numbers, depending on the context.

Our first order of business is to define the notion of 'convergence' for a sequence of test functions. Recall that $\varphi(x)$ is a *test function* on $\mathbb{R}^1$ if:

(a) $\varphi(x)$ is C^∞ (infinitely differentiable);
(b) $\varphi(x)$ has compact support (i.e. $\varphi(x)$ vanishes outside of some compact interval $[a, b]$).

The definition of convergence is motivated – in a sense almost forced – by the restrictions imposed on the test functions themselves.

Definition. A sequence $\varphi_n(x)$ of test functions *converges to zero* (write $\varphi_n \to 0$) if:

(a) for each k, the sequence of kth derivatives $\varphi_1^{(k)}(x), \varphi_2^{(k)}(x), \ldots$ converges uniformly to zero;
(b) the φ_n have uniformly bounded supports, i.e. there is an interval $[a, b]$, independent of n, such that every $\varphi_n(x)$ vanishes outside of $[a, b]$.

Similarly we say that $\varphi_n \to \varphi$ if the sequence $(\varphi - \varphi_n) \to 0$.

Remarks. This notion of convergence is very 'strong' – it requires uniform convergence (in itself a very strong mode of convergence), and moreover this uniformity is required for each of the infinitely many derivatives of the sequence φ_n. In addition, the definition requires a uniform compact support for the entire sequence. However, there is one degree of flexibility. The convergence of $\varphi_n^{(k)}(x)$ as $n \to \infty$ is *not* uniform in k; it is merely uniform in x for each fixed k.

Finally a word about our treatment of the topology. The standard format for a topological theory begins with a preassigned class of 'open' sets. Then each topological notion – including the convergence of sequences – is defined in terms of these open sets. This is done in more technical treatments of distribution theory. For our purposes, it represents a digression. It introduces a host of questions, some of considerable subtlety, which are not essential at this stage. Later, in Chapter 7, we shall discuss the topology in greater depth; the reader can find complete treatments in many sources, e.g. [SW], [GSV] and [ED].

Now that convergence of test functions has been defined, our development follows the pattern outlined in Chapter 1.

Definition. A *distribution* T is a mapping from the set $\mathscr{D}$ of all test functions into the real or complex numbers, such that the following conditions hold (where we write the action of T on a test function φ as $\langle T, \varphi \rangle$):

(a) (Linearity.) $\langle T, a\varphi(x) + b\psi(x) \rangle = a \cdot \langle T, \varphi(x) \rangle + b \cdot \langle T, \psi(x) \rangle$ for all test functions φ, ψ and all constants a, b.

(b) (Continuity.) If $\varphi_n(x) \to 0$ in the sense defined above, then $\langle T, \varphi_n(x) \rangle \to 0$.

As is standard, we shall denote the set of all test functions by $\mathscr{D}$, and the set of all distributions by $\mathscr{D}'$.

In order to motivate what comes next, we refer to Steps 1–3 in the preceding chapter. Step 1 gave the transition from an ordinary function f to the corresponding distribution T_f. This is embodied in the next definition below. As promised in Chapter 1, we bypass Step 2. Thus we start with Step 3, simply laying down the appropriate formula (whose motivation we discovered in Step 2) as a definition. This is the second definition below. Finally Step 2 shows up implicitly in the proof of the Consistency Theorem.

Definition. Let $f(x)$ be a piecewise continuous function on the real line. Then we define the distribution T_f corresponding to f by

$$\langle T_f, \varphi \rangle = \int_{-\infty}^{\infty} f(x)\varphi(x)\,\mathrm{d}x.$$

Definition. Let S and T be arbitrary distributions. Then we define new distributions $S + T$, aT ($a = $ constant), T', $T(ax)$ ($a \neq 0$ is constant), $T(x - a)$, $g(x)T(x)$ (where $g(x)$ is a C^∞ function) by:

(1) $\langle S + T, \varphi \rangle = \langle S, \varphi \rangle + \langle T, \varphi \rangle$;

(2) $\langle aT, \varphi \rangle = a \cdot \langle T, \varphi \rangle$;

(3) $\langle T', \varphi \rangle = -\langle T, \varphi' \rangle$;

(4) $\langle T(ax), \varphi \rangle = |a|^{-1} \langle T, \varphi(x/a) \rangle$;

(5) $\langle T(x - a), \varphi \rangle = \langle T, \varphi(x + a) \rangle$;

(6) $\langle g(x)T(x), \varphi \rangle = \langle T, g(x)\varphi(x) \rangle$.

These, together with convolution (defined in the second half of this chapter), are the primary operations on distributions. This may seem a rather restrictive list: e.g. we have no definition for the product $S \cdot T$ of two distributions (see, however, Chapter 7). Unfortunately, the price we pay for introducing generalized functions (distributions) is that many operations on ordinary functions make no sense in this wider context. Nevertheless, the theory amply justifies itself, as we shall try to show.

Theorem 2.1. All of the operators defined above are distributions.

Proof. We have to show that the operator T_f, together with those defined in (1)–(6) above, satisfy the conditions (a) and (b) in the definition of a distribution. Condition (a) (linearity) is trivial. Therefore we turn to condition (b) (continuity). We give the proofs for T_f and for the derivative T'; the others are similar.

Consider T_f. We need to show that $\varphi_n \to 0$ implies $\langle T_f, \varphi_n \rangle \to 0$. Since $\varphi_n \to 0$, the functions φ_n are supported on a fixed interval $[a, b]$, and the sequence $\varphi_n(x) \to 0$ uniformly as $n \to \infty$. (Here we need not consider $\varphi_n^{(k)}(x) \to 0$ for $k > 0$.) Now:

$$\langle T_f, \varphi_n \rangle = \int_{-\infty}^{\infty} f(x)\varphi_n(x) \, dx = \int_a^b f(x)\varphi_n(x) \, dx.$$

Since f is piecewise continuous, it is bounded on the compact interval $[a, b]$. Hence, $f(x)\varphi_n(x) \to 0$ uniformly as $n \to \infty$, and thus the integral $\langle T_f, \varphi_n \rangle \to 0$.

Consider T'. Recall that $\langle T', \varphi \rangle = -\langle T, \varphi' \rangle$, and that T itself is a distribution, hence continuous. Now we assert that $\varphi_n \to 0$ implies $\varphi_n' \to 0$: for if $\varphi_n^{(k)}(x) \to 0$ uniformly for all k, then so does the sequence $\varphi_n^{(k+1)}(x)$. Thus $\varphi_n \to 0$ implies $\varphi_n' \to 0$, which implies that $\langle T', \varphi_n \rangle = -\langle T, \varphi_n' \rangle \to 0$.

Q.E.D.

Theorem 2.2 (Consistency Theorem). Let $f(x)$ and $e(x)$ be piecewise continuous functions. A further restriction on f is given in (3) below. Then:

(1) $T_f + T_e = T_{f(x)+e(x)}$;

(2) $a \cdot T_f = T_{af(x)}$ ($a =$ constant);

(3) $(T_f)' = T_{f'(x)}$, where f is continuous and piecewise C^1;

(4) $(T_f)(ax) = T_{f(ax)}$ ($a \neq 0$);

(5) $(T_f)(x - a) = T_{f(x-a)}$;

(6) $g(x) \cdot (T_f) = T_{g(x)f(x)}$.

(Thus in each case the distribution operation on the left is consistent with the ordinary function operation on the right.)

Proof. These proofs were already hinted at in Step 2 in Chapter 1, Section 4. We give proofs for (3) and (4); the others are left to the reader.

Consider $(T_f)'$. By definition:

$$\langle (T_f)', \varphi \rangle = -\langle T_f, \varphi' \rangle = -\int_{-\infty}^{\infty} f(x)\varphi'(x)\,dx.$$

Now, since φ has compact support, we integrate by parts in the expression on the right to get:

$$\int_{-\infty}^{\infty} f'(x)\varphi(x)\,dx = \langle T_{f'(x)}, \varphi \rangle.$$

Actually, a little more care is needed, since we assumed only that $f(x)$ is continuous and piecewise C^1. We must break the support $[a, b]$ of φ into subintervals $[a_i, a_{i+1}]$, where $a = a_0 < a_1 < \cdots < a_n = b$, and the a_i are the discontinuity points of $f'(x)$. Then we integrate by parts over each subinterval $[a_i, a_{i+1}]$. This produces boundary terms $f(a_{i+1})\varphi(a_{i+1}) - f(a_i)\varphi(a_i)$. But since f is continuous, the sum ($i = 0$ to $n - 1$) of the boundary terms cancels out, and the proof goes as before.

Consider $(T_f)(ax)$. By definition:

$$\langle (T_f)(ax), \varphi \rangle = |a|^{-1}\langle T_f, \varphi(x/a) \rangle = |a|^{-1}\int_{-\infty}^{\infty} f(x)\varphi(x/a)\,dx.$$

Now suppose $a > 0$, and make the substitution $u = x/a$, $du = dx/a$. We obtain:

$$\int_{-\infty}^{\infty} f(au)\varphi(u)\,du = \langle T_{f(ax)}, \varphi \rangle.$$

If $a < 0$, then the substitution $u = x/a$ leads to $\int_{\infty}^{-\infty}$, in which the limits of integration are in the wrong order. Reversing the limits of integration introduces a factor of (-1), which explains the factor $|a|^{-1}$ in the original definition. Q.E.D.

We introduce one more definition, that of convergence for a sequence of distributions.

Definition. Let $\{T_n\}$ be a sequence of distributions and let T be a distribution. We say that T_n *converges* to T, written $T_n \to T$, if

$$\langle T_n, \varphi \rangle \to \langle T, \varphi \rangle \text{ for every test function } \varphi.$$

Remarks. In functional analysis, such a mode of convergence is traditionally called 'weak' or 'weak-star' convergence. In this case it is very weak indeed. We remarked earlier that our definition of convergence for test functions is very 'strong'. Convergence of distributions is 'weak', i.e. easy to satisfy, precisely because the conditions we impose on the test functions are so stringent. This weakness of convergence is one of the most pleasing aspects of the theory, for it allows us to prove convergence under a wide range of circumstances (see the Examples below).

The assumption made above, that the limit T is a distribution, is actually unnecessary. This can be proved by using the 'Banach–Steinhaus Theorem' for topological vector spaces. Because of our presentation, we must omit the proof. We shall, in fact, have no reason to use this result: in every case we consider the fact that T is a distribution will be clear on other grounds.

Theorem 2.3 (continuity of the distribution operations). Let $\{S_n\}$ and $\{T_n\}$ be sequences of distributions converging to distributions S and T, respectively. Then:

(1) $(S_n + T_n) \to S + T$;

(2) $a \cdot T_n \to aT$;

(3) $T_n' \to T'$;

(4) $T_n(ax) \to T(ax)$;

(5) $T_n(x - a) \to T(x - a)$;

(6) $g(x) \cdot T_n(x) \to g(x)T(x)$.

Proof. We give the proof for (3); the others are similar. Let $T_n \to T$. We want to show that $T_n' \to T'$, i.e. that $\langle T_n', \varphi \rangle \to \langle T', \varphi \rangle$ for every test function φ. But $\langle T_n', \varphi \rangle = -\langle T_n, \varphi' \rangle$ and $\langle T', \varphi \rangle = -\langle T, \varphi' \rangle$. Since φ' is also a test function, and $T_n \to T$, we have $\langle T_n, \varphi' \rangle \to \langle T, \varphi' \rangle$, as desired.   Q.E.D.

The above result is far more remarkable than its proof. It has no analog in classical function theory. For example, the sequence of functions $\{n^{-1} \sin nx\}$ converges uniformly to zero, but its sequence of derivatives $\{\cos nx\}$ oscillates. More precisely, the derivatives diverge when viewed in the classical 'pointwise' sense. When viewed in the distribution sense, the sequence of derivatives converges to zero because the original sequence did. Similarly, by using the $(k + 1)$st derivative, we see that the sequences $\{n^k \cos nx\}$ and $\{n^k \sin nx\}$ converge to zero in $\mathcal{D}'$ as $n \to \infty$, for any fixed power k. Further examples illustrating this theme will be given below.

Theorem 2.4 (identities of calculus). Let S and T be distributions, $g(x)$ a C^∞ function, and $a \neq 0$ a real constant. Then:

(1) $(S + T)' = S' + T'$;

(2) $(aT)' = a \cdot T'$;

(3) (the third definition was of T' itself);

(4) $[T(ax)]' = a \cdot T'(ax)$;

(5) $[T(x-a)]' = T'(x-a)$;

(6) $[g(x) \cdot T(x)]' = g'(x)T(x) + g(x)T'(x)$.

Proof. These proofs are amusing, because of the way that the duality in the definition of $\langle T, \varphi \rangle$ juggles the calculus operations around. We shall do (6). Readers wanting a little bit of practice might want to do (4). For (6), we take the (possibly unexpected) starting point $g(x)T'(x)$:

$$\langle gT', \varphi \rangle = \langle T', g\varphi \rangle = -\langle T, (g\varphi)' \rangle$$

$$= -\langle T, g'\varphi + g\varphi' \rangle = -\langle T, g'\varphi \rangle - \langle T, g\varphi' \rangle$$

$$= -\langle g'T, \varphi \rangle - \langle gT, \varphi' \rangle = -\langle g'T, \varphi \rangle + \langle (gT)', \varphi \rangle.$$

Thus $gT' = -g'T + (gT)'$, which yields $(gT)' = g'T + gT'$, as desired.

Q.E.D.

We now illustrate the theory with some examples.

Example 1 (the delta function). This is defined by the equation

$$\langle \delta, \varphi \rangle = \varphi(0).$$

Now suppose we translate the delta function through a distance $+a$. This should give the function $\delta(x-a)$, and we find

$$\langle \delta(x-a), \varphi \rangle = \langle \delta, \varphi(x+a) \rangle = \varphi(a).$$

Thus the translate $\delta(x-a)$ evaluates the test function φ at $x = a$, just as we expected. The following may be a little more surprising:

$$\delta(2x) = (1/2) \cdot \delta(x).$$

For $\langle \delta(2x), \varphi \rangle = 2^{-1}\langle \delta, \varphi(x/2) \rangle = 2^{-1} \cdot \varphi(0) = 2^{-1}\langle \delta, \varphi \rangle$. More generally, of course, $\delta(ax) = |a|^{-1} \delta(x)$. In particular, $\delta(-x) = \delta(x)$.

Theorem 2.5 (approximate identities). Let $f(x)$ be a piecewise continuous function such that $\int_{-\infty}^{\infty} |f(x)| \, dx < \infty$ and $\int_{-\infty}^{\infty} f(x) \, dx = 1$. We write $f_a(x) = af(ax)$. Then,

$$f_a(x) = af(ax) \to \delta(x) \quad \text{as } a \to \infty.$$

Proof. Of course, the above convergence is in the sense of distributions. The idea is that, as $a \to \infty$, $f_a(x)$ becomes a 'narrow pulse', with its height growing at the same rate that its width shrinks; the integral of f_a remains equal to 1. Here are the details.

Take any test function φ, and take $\varepsilon > 0$. Choose $\delta > 0$ so that $|\varphi(x) - \varphi(0)| < \varepsilon$ for $|x| < \delta$. Choose M so that $(\int_{-\infty}^{-M} + \int_{M}^{\infty})|f(x)| \, dx < \varepsilon$. Finally, take any $a > M/\delta$. Let 'Const' denote the fixed value $\int_{-\infty}^{\infty} |f(x)| \, dx$, and let 'Konst' denote $\max\limits_{x} |\varphi(x)|$. Then,

$$|\langle T_{af(ax)}, \varphi\rangle - \langle\delta, \varphi\rangle| = \left|\int_{-\infty}^{\infty} af(ax)\varphi(x)\,dx - \varphi(0)\right|$$

$$= \left|\int_{-\infty}^{\infty} f(u)[\varphi(u/a) - \varphi(0)]\,du\right|$$

(where we set $u = ax$ and use $\int_{-\infty}^{\infty} f = 1$)

$$\leqslant \left|\int_{-M}^{M} f(u)[\varphi(u/a) - \varphi(0)]\,du\right| + \left|\int_{|u|>M} f(u)[\varphi(u/a) - \varphi(0)]\,du\right|.$$

The first term in the last line above we dominate by

$$\left(\int_{-\infty}^{\infty} |f(u)|\,du\right)\cdot\left(\max_{|u|\leqslant M} |\varphi(u/a) - \varphi(0)|\right).$$

Now, since $|u| \leqslant M$ and $a > M/\delta$, $|u/a| < \delta$. Thus the product above is $\leqslant \text{Const}\cdot\varepsilon$. The second term we dominate by

$$\left(\int_{|u|>M} |f(u)|\,du\right)\cdot\left(\max_{u} |\varphi(u/a) - \varphi(0)|\right),$$

which is $\leqslant \varepsilon\cdot(2\cdot\text{Konst})$, by definition of M and 'Konst'. Thus the total error is dominated by

$$\text{Const}\cdot\varepsilon + 2\cdot\text{Konst}\cdot\varepsilon.$$

Since Const and Konst are fixed, and we can make ε arbitrarily small, this completes the proof.

An important special case of Theorem 2.5 occurs when $f(x)$ is the Gaussian (or normal) density function

$$f(x) = (2\pi)^{-1/2}\,e^{-x^2/2}.$$

Then as $a \to \infty$, $f_a(x) = af(ax) \to \delta(x)$, $f'_a \to \delta'$, $f''_a \to \delta''$, etc. Thus the derivatives of the delta function appear as the limits, in the distribution sense, of standard functions. The following example was presented without proof in Chapter 1:

$$\sum_{n=-\infty}^{\infty} \delta''(x - 2\pi n) = -\frac{1}{\pi}\sum_{n=1}^{\infty} n^2\cdot\cos nx.$$

To prove it, we simply take the well known Fourier series

$$2\cdot\sum_{n=1}^{\infty} \frac{\sin nx}{n} = \begin{cases} \pi - x & \text{for } 0 < x \leqslant \pi \\ -\pi - x & \text{for } -\pi \leqslant x < 0, \end{cases}$$

where the function on the right is extended periodically: recall that this gives a 'sawtooth' function consisting of disconnected straight lines of slope -1, with jumps of 2π at the points $2\pi n$, $n = 0, \pm 1, \pm 2, \ldots$. Now by Theorem

2.3 we can differentiate term by term to obtain

$$\frac{1}{2\pi} + \frac{1}{\pi}\sum_{n=1}^{\infty}\cos nx = \sum_{n=-\infty}^{\infty}\delta(x - 2\pi n),$$

since, after dividing by 2π, the derivatives of the 'jumps' produce the Delta Functions, and the slopes $= -1$ show up as the term $1/2\pi$. Now differentiating two more times we get

$$-\frac{1}{\pi}\sum_{n=1}^{\infty}n^2 \cdot \cos nx = \sum_{n=-\infty}^{\infty}\delta''(x - 2\pi n).$$

We remark that the first formula, for $\Sigma\,\delta(x - 2\pi n)$, corresponds to the Dirichlet kernel

$$D_N(x) = \frac{1}{2\pi} + \frac{1}{\pi}\sum_{n=1}^{N}\cos nx,$$

where we have let $N \to \infty$. The fact that the limit is a periodic sequence of delta functions gives an abstract analog of Dirichlet's proof of the convergence of Fourier series. (To establish the connection, we would need convolutions, defined in Part 2 of this chapter. We shall not return to this theme, since the classical proofs give considerably more information, but we thought the analogy deserved mention.)

Theorem 2.6. Let T be any distribution. Then (with convergence in the sense of distributions),

$$T'(x) = \lim_{h \to 0}\frac{T(x + h) - T(x)}{h}.$$

Proof. By definition, for any test function φ, $\langle T', \varphi \rangle = -\langle T, \varphi' \rangle$. Similarly $\langle [T(x + h) - T(x)]/h, \varphi \rangle = \langle T, [\varphi(x - h) - \varphi(x)]/h \rangle$. Now, as $h \to 0$, $[\varphi(x - h) - \varphi(x)]/h \to -\varphi'(x)$ in $\mathscr{D}$. Since T is a continuous operator on $\mathscr{D}$, the result follows.

Remark. This is the first time we have used the fact that distributions are continuous. (Theorem 2.1, where we use continuity to prove continuity, does not count. Theorem 2.3 did *not* depend on the continuity of the distributions themselves.) However, we will find continuity being used extensively in the second part of this chapter.

Part 2 Convolutions

This part of Chapter 2 is a little more technical than the preceding one and could be omitted on a first reading.

Translation of a function. If $f(x)$ is a function and a is a real number, then $f(x - a)$ is called the *translate* of $f(x)$ through the distance $+a$. Thus the graph of $f(x-a)$ coincides with the graph of $f(x)$ moved horizontally by a distance a; for example, the position $x = 0$ for $f(x)$ corresponds to the position $x = a$ for $f(x-a)$ (see Figure 6).

1 Convolution of test functions

The convolution $\varphi * \psi$ of two test functions φ and ψ is defined by

$$(\varphi * \psi)(x) = \int_{-\infty}^{\infty} \varphi(x-y)\psi(y)\,dy.$$

Intuitively we may think of this as a 'generalized sum' (i.e. an integral) of translates $\varphi(x-y)$ of $\varphi(x)$. In this process, the translate $\varphi(x-y)$ is multiplied by the 'weight factor' $\psi(y)$, and then these translates are averaged by integrating with respect to y.

Among the many applications of this concept, we mention here its connection with probability theory. Suppose we have two independent experiments whose outputs are real numbers, varying according to some probability law. (In probability theory these are called 'random variables'.) Suppose that the outputs of these experiments are distributed along the real line with continuous probability densities $\varphi(x)$ and $\psi(x)$, respectively. Then the *sum* of the two outputs will have the probability density $(\varphi * \psi)(x)$, i.e. the probability for the sum is the convolution of the two original probabilities.

Since this is not a text on probability, we have no need to prove this rigorously – but it is easy to see intuitively why it is true. We ask: How can the sum of the two experiments produce any given value x? Well, the second experiment (corresponding to ψ), could take any value y, provided that the first experiment (corresponding to φ) takes the complementary value $x-y$. Since the experiments are independent, the probability densities $\varphi(x-y)$ and $\psi(y)$ should be multiplied, and then these products should be 'added' (integrated) over all possible values of y. Of course this gives the convolution integral above.

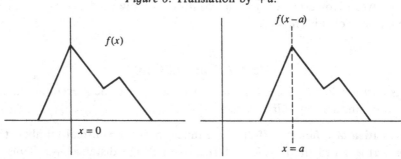

Figure 6. Translation by $+a$.

We now list some of the basic properties of convolution. The proofs will follow. First we have the commutative and associative laws:

$$\varphi * \psi = \psi * \varphi$$

$$(\varphi * \psi) * \tau = \varphi * (\psi * \tau).$$

Since the convolution is an 'average' of translates $\varphi(x - y)$, it is not surprising that convolution commutes with translation and with differentiation, i.e.

$$(\varphi * \psi)(x - a) = \varphi(x - a) * \psi(x) = \varphi(x) * \psi(x - a),$$

$$(d/dx)(\varphi * \psi) = \varphi' * \psi = \varphi * \psi'.$$

Finally, the obvious fact that $\tau * (\varphi + \psi) = (\tau * \varphi) + (\tau * \psi)$, i.e. distributivity over addition, does not deserve to be displayed.

Proof of the convolution identities. Since the test functions are C^∞ with compact support, we can apply all of the usual rules of calculus. By doing the computations we find.

For $\varphi * \psi = \psi * \varphi$:

$$(\varphi * \psi)(x) = \int_{-\infty}^{\infty} \varphi(x - y)\psi(y)\,dy$$

and setting $u = x - y$, $du = -dy$ gives

$$-\int_{\infty}^{-\infty} \varphi(u)\psi(x - u)\,du = \int_{-\infty}^{\infty} \psi(x - u)\varphi(u)\,du.$$

The associative law $(\varphi * \psi) * \tau = \varphi * (\psi * \tau)$ is similar. First we use the commutativity, already proved, to rewrite the desired formula as $(\psi * \varphi) * \tau = (\psi * \tau) * \varphi$. Now

$$((\psi * \varphi) * \tau)(x) = \int_{-\infty}^{\infty} (\psi * \varphi)(x - y)\tau(y)\,dy$$

$$= \int_{-\infty}^{\infty} \int_{-\infty}^{\infty} \psi(x - y - z)\varphi(z)\tau(y)\,dz\,dy$$

$$= \int_{-\infty}^{\infty} \varphi(z) \int_{-\infty}^{\infty} \psi(x - z - y)\tau(y)\,dy\,dz$$

$$= ((\psi * \tau) * \varphi)(x).$$

The formula for $(\varphi * \psi)(x - a)$ is trivial. Finally, to compute $(d/dx)(\varphi * \psi)$, we differentiate under the integral

$$(d/dx) \int_{-\infty}^{\infty} \varphi(x - y)\psi(y)\,dy = \int_{-\infty}^{\infty} \varphi'(x - y)\psi(y)\,dy.$$

Thus $(d/dx)(\varphi * \psi) = \varphi' * \psi$, and since we have already proved commutativity this must equal $\varphi * \psi'$ as well.

Note. It appears from the formula $(\varphi * \psi)' = \varphi' * \psi$, that, in order to insure the differentiability of $\varphi * \psi$, it suffices that just one of the two functions φ

or ψ be differentiable. This is true; the convolution is an averaging process, and the convolution of a 'jagged' function f by a smooth function φ produces a smooth function as the output. In fact, for this reason, convolutions are sometimes called 'smoothing operator'. This idea will have important applications in the next section.

Now we come to some more technical facts which are useful in distribution theory.

It is easily verified that, if the test function $\varphi(x)$ has support on $[a, b]$ and if $\psi(x)$ has support on $[c, d]$, then the convolution $(\varphi * \psi)(x)$ has support on $[a + c, b + d]$. Thus convolution preserves the property of having compact support. From the formula $(\varphi * \psi)' = \varphi' * \psi$, which gives $(\varphi * \psi)^{(k)} = \varphi^{(k)} * \psi$, we see that convolution also preserves the property of being C^∞. Thus we have the important conclusion: *the convolution of two test functions is a test function.*

We conclude this section with an identity which will play the role of 'Step 2' when we extend these definitions to distributions. First we recall that a function $f(x)$ acts on a test function $\psi(x)$ to produce the distribution

$$\langle f, \psi \rangle = \int_{-\infty}^{\infty} f(x)\psi(x) \, dx.$$

When f itself is a test function, we have the following additional identities. We define

$$\tilde{\psi}(x) = \psi(-x).$$

Then, for any test function φ, ψ and τ we have

$$\psi^{\sim\sim} = \psi$$
$$\langle \varphi, \psi \rangle = (\varphi * \tilde{\psi})(0)$$
$$(\varphi * \psi)^{\sim} = \tilde{\varphi} * \tilde{\psi}$$
$$\langle \varphi * \tau, \psi \rangle = \langle \varphi, \tilde{\tau} * \psi \rangle.$$

We remark that the identity $\langle \varphi, \psi \rangle = (\varphi * \tilde{\psi})(0)$ allows us to express the inner product as a convolution. The identity $\langle \varphi * \tau, \psi \rangle = \langle \varphi, \tilde{\tau} * \psi \rangle$ is particularly important. This is because eventually we shall replace φ and τ by distributions, although ψ will still be a test function. We will define the convolution of distributions in the usual way: by moving the operation on the distributions over to a corresponding operation on the test function. The last identity shows us how to do this.

Proofs of these identities. The first three formulas are trivial. For the last one we use the associative law:

$$\langle \varphi * \tau, \psi \rangle = [(\varphi * \tau) * \tilde{\psi}](0)$$
$$= [\varphi * (\tau * \tilde{\psi})](0)$$
$$= [\varphi * (\tilde{\tau} * \psi)^{\sim}](0)$$
$$= \langle \varphi, \tilde{\tau} * \psi \rangle.$$

2 Convolution of a distribution and a test function

This asymmetrical situation serves as an intermediate step towards the definition of convolution for two distributions. It also leads to an important approximation theorem.

Before we proceed to our main topic, we must introduce the notion of compact support for a distribution. Recall that the support of a function $f(x)$ is the closure of the set of points x where $f(x) \neq 0$. For example, the test functions, because they vanish outside of some bounded set, have compact support. Now, defining the support of a distribution is rather tricky, and we prefer to omit it. We are not interested in describing *the* support (i.e. the smallest support) but only the notion of 'compact' or 'bounded' support. This is very simple.

Definition. A distribution T has *compact support* if there is a compact set $[a, b]$ such that, for all test functions φ whose support lies outside of $[a, b]$, $\langle T, \varphi \rangle = 0$. In this case we also say that the support of T lies *within* $[a, b]$, written $\text{support}(T) \subseteq [a, b]$.

Definition. Let T be a distribution and φ a test function. We define $T * \varphi$ by

$$(T * \varphi)(x) = \langle T(y), \varphi(x - y) \rangle = \langle T(x - y), \varphi(y) \rangle.$$

Note that $T * \varphi$ is a *bona fide* function: we define $(T * \varphi)(x)$ to be the number $\langle T(y), \varphi(x - y) \rangle$, i.e. the distribution $T(y)$ acting on the test function $\varphi_x(y) = \varphi(x - y)$. With the above definition, the inner product $\langle T, \varphi \rangle$ can be expressed as a convolution for any distribution T:

$$\langle T, \varphi \rangle = (T * \tilde{\varphi})(0).$$

We also observe that this is consistent with our previous definition in the case where $T = T_\psi$ is a test function $\psi(y)$:

$$(T_\psi * \varphi)(x) = \langle T_\psi(y), \varphi(x - y) \rangle = \int_{-\infty}^{\infty} \varphi(x - y) \psi(y) \, \mathrm{d}y.$$

Theorem 2.7. $T * \varphi$ is a C^∞ function. Furthermore, if T has compact support, then so does $T * \varphi$, i.e. $T * \varphi$ is a test function.

To prove Theorem 2.7 we need the following lemmas which are of importance in themselves.

Lemma 2.8. $T * \varphi$ is continuous.

Proof. Since φ is a test function, φ and its derivatives are uniformly continuous. Thus, for each k, $\varphi^{(k)}(z - y) \to \varphi^{(k)}(x - y)$ uniformly as $z \to x$, i.e.

$\varphi(z - y) \to \varphi(x - y)$ in the sense of test functions. Therefore, by continuity of T, $\langle T(y), \varphi(z - y) \rangle \to \langle T(y), \varphi(x - y) \rangle$ as $z \to x$. Hence $T * \varphi$ is continuous.

Lemma 2.9. $(T * \varphi)' = T * \varphi'$.

Proof. In the first part of Chapter 2 we showed that

$$T'(x) = \lim_{h \to 0} [T(x + h) - T(x)]/h.$$

Now $(T * \varphi)'(x)$ is

$$\lim_{h \to 0} [\langle T(x + h - y), \varphi(y) \rangle - \langle T(x - y), \varphi(y) \rangle]/h,$$

whereas

$$(T * \varphi')(x) = \langle T(x - y), \varphi'(y) \rangle = \langle T'(x - y), \varphi(y) \rangle,$$

since the $' = d/dy$.

Lemma 2.10. If support$(T) \subseteq [-a, a]$ and support$(\varphi) \subseteq [-b, b]$, then support$(T * \varphi) \subseteq [-a - b, a + b]$.

Proof. As a function of y, support$(\varphi(x - y)) \subseteq [-b + x, b + x]$. If $b + x < -a$ or $a < -b + x$, i.e. $x \notin [-a - b, a + b]$, then support$(\varphi(x - y))$ lies outside of $[-a, a]$ and $(T * \varphi)(x) = \langle T(y), \varphi(x - y) \rangle = 0$. Hence support$(T * \varphi) \subseteq [-a - b, a + b]$.

Proof of Theorem 2.7. Obvious from the above lemmas.   Q.E.D.

The following two theorems will lead us to our final goal, which is the convolution of distributions.

Theorem 2.11. Let T be a distribution and let φ and ψ be test functions. Then

$$(T * \varphi) * \psi = T * (\varphi * \psi).$$

Proof. $(\varphi * \psi)(x) = \int_{-\infty}^{\infty} \varphi(x - y)\psi(y) \, dy$ is the limit of the Riemann sum $r_n(x) = (1/n) \sum_{m=-\infty}^{\infty} \varphi[x - (m/n)]\psi(m/n)$. (Since φ and ψ have compact support, this is a finite sum.) Indeed, $r_n(x) \to (\varphi * \psi)(x)$ uniformly as $n \to \infty$ because φ, ψ and $\varphi * \psi$ are uniformly continuous with compact support. Likewise, for each k, $r_n^{(k)} \to (\varphi * \psi)^{(k)}$ uniformly since $(\varphi * \psi)^{(k)} = \varphi^{(k)} * \psi$. Hence, $r_n \to \varphi * \psi$ in $\mathscr{D}$. Thus we have

$$(T * (\varphi * \psi))(x) = \langle T(y), (\varphi * \psi)(x - y) \rangle$$

$$= \langle T(y), \lim_{n \to \infty} r_n(x - y) \rangle$$

$$= \lim_{n \to \infty} \langle T(y), r_n(x - y) \rangle \quad \text{(by continuity of } T\text{)}$$

$$= \lim_{n \to \infty} 1/n \sum_{m=-\infty}^{\infty} \left\langle T(y), \varphi\left(x - y - \frac{m}{n}\right)\psi\left(\frac{m}{n}\right)\right\rangle \quad \text{(by linearity of } T)$$

$$= \int_{-\infty}^{\infty} \langle T(y), \varphi(x - y - z)\rangle \psi(z)\, dz$$

$$= \int_{-\infty}^{\infty} (T * \varphi)(x - z)\psi(z)\, dz$$

$$= ((T * \varphi) * \psi)(x). \hspace{4cm} \text{Q.E.D.}$$

The next theorem provides the key to removing the 'topological' difficulties in our treatment of convolution.

Theorem 2.12. Let T have compact support. If $\varphi_n \to \varphi$ in $\mathscr{D}$, then $T * \varphi_n \to T * \varphi$ in $\mathscr{D}$.

Proof. Since T has compact support, and the φ_n have uniformly bounded support, we know by Lemma 2.10 that the convolutions $T * \varphi_n$ have uniformly bounded support. Let $[-c, c]$ be an interval which contains the support of $T * \varphi_n$ for all n.

Now it would be easy to show that the sequence $(T * \varphi_n)(x)$ converges 'pointwise' to $(T * \varphi)(x)$, i.e. that it converges for each particular point x. However, we need *uniform* convergence, and this requires a little more work. We use a topological trick.

Let K be the compact set in $\mathbb{R}^2$ consisting of all points (x, s) where $-c \leqslant x \leqslant c$ and s runs through the values $1, 1/2, 1/3, \ldots, 1/n, \ldots$ and the limiting value 0. We identify the nth term in the sequence $(T * \varphi_n)(x)$ with the value $s = 1/n$, and the limit $(T * \varphi)(x)$ with $s = 0$. Our proof will be based on the fact that a continuous function on a compact set is uniformly continuous.

Let $F(x, s) = (T * \varphi_n)(x)$ for $s = 1/n$, $F(x, 0) = (T * \varphi)(x)$. We need to show that $F(x, s)$ is continuous in both variables; the only difficulty occurs when $s = 0$.

Lemma. If $\varphi_n \to \varphi$ in the test function sense, and $x_n \to x$ in $\mathbb{R}$, then the functions (of y)

$$\varphi_n(x_n - y) \to \varphi(x - y)$$

in $\mathscr{D}$.

Proof. Since $\varphi_n \to \varphi$ uniformly and φ is uniformly continuous, $\varphi_n(x_n - y) \to \varphi(x - y)$ uniformly in y. We can say the same for the derivatives because $\varphi_n \to \varphi$ in $\mathscr{D}$. Thus $\varphi_n(x_n - y) \to \varphi(x - y)$ in $\mathscr{D}$. \hspace{1cm} Q.E.D.

Of course we could replace the sequence of indices $n = 1, 2, 3, \ldots$ by any other sequence of integers n_k approaching infinity. Thus, combining the lemma with the continuity of the distribution T, we see that $F(x, s)$ is continuous *on* K. Hence it is also uniformly continuous. Consequently $F(x, 1/n) = (T * \varphi_n)(x)$ converges uniformly as $n \to \infty$ to $F(x, 0) = (T * \varphi)(x)$. Since $(T * \varphi_n)' = T * \varphi_n'$ and $(T * \varphi)' = T * \varphi'$, the same thing holds for the first derivatives; similarly for each higher derivative. Thus, for each k, $(T * \varphi_n)^{(k)}(x)$ converges uniformly as $n \to \infty$ to $(T * \varphi)^{(k)}(x)$. Since all of the supports are contained in $[-c, c]$, this shows that $(T * \varphi_n) \to (T * \varphi)$ in $\mathscr{D}$.

Q.E.D.

Approximate identities. One of the main applications of convolutions is that they provide a coherent method for generating approximations.

Theorem 2.13. Let θ be a test function such that

$$\theta(x) \geqslant 0 \qquad \text{and} \qquad \int_{-\infty}^{\infty} \theta(x)\, dx = 1.$$

Let $\theta_a(x) = a\theta(ax)$, $a > 0$. Then, for any test function ψ,

$$\lim_{a \to \infty} \theta_a * \psi = \psi,$$

the convergence being in the sense defined for test functions.

Proof. The idea behind the proof is that, as $a \to \infty$, the pulse θ_a becomes very narrow and very high, while its total area remains equal to one. Thus, as $a \to \infty$, θ_a behaves like an approximation to the Dirac Delta Function. This is the genesis of the term 'approximate identity'. Now for the details.

Since $(\theta_a * \psi)^{(k)} = \theta_a * \psi^{(k)}$ and the $\psi^{(k)}$ are test functions themselves, it suffices to show that $\theta_a * \psi \to \psi$ uniformly.

Let $b > 0$ be such that $\operatorname{support}(\theta) \subseteq [-b, b]$. Then $\operatorname{support}(\theta_a) \subseteq [-b/a, b/a]$ and

$$|(\theta_a * \psi)(x) - \psi(x)| = \left| \int_{-b/a}^{b/a} \theta_a(y)\psi(x - y)\, dy - \int_{-b/a}^{b/a} \theta_a(y)\psi(x)\, dy \right|$$

$$\leqslant \int_{-b/a}^{b/a} \theta_a(y)|\psi(x - y) - \psi(x)|\, dy.$$

Here we have used the fact that $\int \theta_a(y)\, dy = 1$. Now the last integral approaches zero uniformly as $a \to \infty$ because ψ is uniformly continuous.

Q.E.D.

Now we use convolution with an approximate identity to prove one of the most striking results in the whole theory. Suppose we arrange functions and distributions in a hierarchy according to their smoothness. The smoothest would be the C^∞ functions, then C^n, then continuous, then functions which are not continuous, and finally the distributions which are so 'wild' that they do not correspond to functions at all (e.g. the Delta Function). Now it turns out that every distribution, no matter how wild, can be approximated by a sequence of C^∞ functions.

Theorem 2.14. Every distribution T is the limit in the distribution sense of a sequence φ_n of C^∞ functions. If T has compact support, then the φ_n will be test functions. Furthermore the φ_n can be chosen to satisfy the following stronger condition.

For any test function ψ,

$$(\varphi_n * \psi)(x) \to (T * \psi)(x)$$

in $\mathscr{D}$ (i.e. the convergence is uniform in x for the functions and each of their derivatives).

Proof. Let $\{\theta_a\}$ be an approximate identity, and let $a \to \infty$ through the values $a = n = 1, 2, 3, \ldots$. Let $\varphi_n = T * \tilde{\theta}_n$. Then φ_n *is* C^∞, and for any test function ψ we have

$$\langle \varphi_n, \psi \rangle = \langle T * \tilde{\theta}_n, \psi \rangle = \langle T, \theta_n * \psi \rangle.$$

These expressions approach $\langle T, \psi \rangle$ because, $\{\theta_n\}$ being an approximate identity, $\theta_n * \psi \to \psi$ in $\mathscr{D}$. If T has compact support, then the φ_n are test functions from Theorem 2.7, and $\varphi_n * \psi = (T * \tilde{\theta}_n) * \psi = T * (\tilde{\theta}_n * \psi)$. Now $\tilde{\theta}_n * \psi \to \psi$ because $\{\tilde{\theta}_n\}$ is also an approximate identity. Hence from Theorem 2.12 we have $\varphi_n * \psi = T * (\tilde{\theta}_n * \psi) \to T * \psi$ in $\mathscr{D}$.   Q.E.D.

3 Convolution of distributions

As suggested by the preceding section, we will define the convolution of an arbitrary distribution with a distribution which has compact support. This case is simple enough to be considered in this chapter.

Remarks. The restriction to compact supports is obviously a nuisance. It is mandated by the fact that, in the second definition of the previous section, the convolution $T * \varphi$ of a distribution T with a test function φ need not have compact support. To obtain compact support for $T * \varphi$ (i.e. to guarantee that $T * \varphi$ is a test function), we need a compact support hypothesis on T.

An extension to a more general situation will be given in Chapter 7. Even this extension is not as general as we might wish: so far, no one has discovered a definition of convolution which is applicable to *all* distributions. In Chapter 7 we will give an example which suggests that probably there is none.

Definition. Let S and T be distributions and suppose that T has compact support. Then we define

$$\langle S * T, \varphi \rangle = \langle S, \tilde{T} * \varphi \rangle, \qquad \text{where } \tilde{T}(x) = T(-x).$$

The consistency of this definition with the one in which T is a test function follows by the identity $\langle S, \varphi \rangle = (S * \tilde{\varphi})(0)$ and Theorem 2.11. The continuity of $S * T$ follows from Theorem 2.12, and hence $S * T$ is a distribution.

For the remainder of this section, in order to achieve symmetry, *we will consider only cases in which both distributions have compact support.*

Theorem 2.15 (Continuity Theorem). If $T_n \to T$ in the sense of distributions, then, for any fixed S, $T_n * S \to T * S$.

Proof. For any test function φ, $\langle T_n * S, \varphi \rangle = \langle T_n, \tilde{S} * \varphi \rangle \to \langle T, \tilde{S} * \varphi \rangle$ because $T_n \to T$ in the sense of distributions and $\tilde{S} * \varphi$ is a test function. Hence $T_n * S \to T * S$.

Remarks. The proof that $S * T_n \to S * T$ is much harder, and we postpone it. Continuity in both variables fails: it is not true that $S_n \to S$ and $T_n \to T$ implies $S_n * T_n \to S * T$. Here is a counterexample: let $S_n = \delta(x - n)$ and $T_n = \delta(x + n)$. Then $S_n \to 0$ and $T_n \to 0$, but $S_n * T_n = \delta(x)$ for all n.

Lemma 2.16. Suppose $T_n \to T$ in the stronger sense suggested by Theorem 2.14; i.e., for each test function ψ, $(T_n * \psi)(x) \to (T * \psi)(x)$ in $\mathscr{D}$. Then $S * T_n \to S * T$ in the sense of distributions.

Proof. $\langle S * T_n, \psi \rangle = \langle S, \tilde{T}_n * \psi \rangle \to \langle S, \tilde{T} * \psi \rangle = \langle S * T, \psi \rangle$ since $(\tilde{T}_n * \psi) \to (\tilde{T} * \psi)$ in $\mathscr{D}$.

Now we are in a position to give a very neat proof of all of the standard convolution identities for distributions with compact support.

Theorem 2.17. Let S, T, U be distributions with compact support. Then,

$$S * T = T * S,$$
$$(S * T) * U = S * (T * U),$$
$$(S * T)' = S * T' = S' * T,$$
$$(S * T)(x - a) = S(x - a) * T(x).$$

Proof. We give the proof for $S * T = T * S$; the other parts are similar. From Theorem 2.14, we know that there exist sequences of test functions $\varphi_n \to S$ and $\tau_m \to T$ which converge in the stronger sense suggested by Theorem 2.14. Since φ_n and τ_m are test functions,

$$\varphi_n * \tau_m = \tau_m * \varphi_n.$$

Now let $m \to \infty$: by Lemma 2.16 and Theorem 2.15, respectively, we know that $\varphi_n * \tau_m \to \varphi_n * T$ and $\tau_m * \varphi_n \to T * \varphi_n$. Hence,

$$\varphi_n * T = T * \varphi_n.$$

Now let $n \to \infty$, and by the same argument we see that $S * T = T * S$.

Remark. Now that we have proved $S * T = T * S$, we can give the other half of the Continuity Theorem without the extra hypothesis used in Lemma 2.16:

$$\text{if } T_n \to T, \text{ then } S * T_n \to S * T.$$

Examples and applications. To give substance to our abstract development, we wish to emphasize applications which require distribution theory – i.e. applications which could not be phrased in terms of ordinary functions. The single most important distribution is the Dirac Delta 'Function'. As we have noted, it is not a *bona fide* function. Recall its definition:

$\langle \delta, \varphi \rangle = \varphi(0)$, i.e. δ maps each test function φ into its value $\varphi(0)$ at $x = 0$. Similarly, as noted in Part 1 of this chapter, $\delta(x - a)$ maps each test function φ into $\varphi(a)$. Recall that intuitively we think of $\delta(x - a)$ as a unit 'impulse' or 'point mass' located at $x = a$. Now the Delta Function acts as an identity under convolution.

Lemma 2.18. For any distribution T with compact support,

$$\delta * T = T * \delta = T.$$

Proof. By commutativity, it suffices to consider only $T * \delta$. By definition:

$$\langle T * \delta, \varphi \rangle = \langle T, \tilde{\delta} * \varphi \rangle.$$

Now $\tilde{\delta}(x) = \delta(-x)$ and

$$(\tilde{\delta} * \varphi)(x) = \langle \tilde{\delta}(x - y), \varphi(y) \rangle = \langle \delta(y - x), \varphi(y) \rangle = \varphi(x).$$

Hence $\langle T * \delta, \varphi \rangle = \langle T, \varphi \rangle$, i.e. $T * \delta = T$.

This lemma has the important consequence that differentiation and translation are both special cases of convolution. Thus convolution emerges as the operation *par excellence* in distribution theory. Its only defect is the restriction to compact supports. This defect will be largely remedied in Chapter 7.

Theorem 2.19. Let T be a distribution with compact support. Then,

$$T'(x) = \delta'(x) * T(x)$$
$$T(x - a) = \delta(x - a) * T(x).$$

Proof. This is an immediate consequence of Theorem 2.17 and Lemma 2.18. Thus, for T',

$$T = \delta * T$$

(by the lemma), and
$$T' = (\delta * T)' = \delta' * T$$
(by the theorem). Similarly for $T(x - a)$.

Corollary. For any integer k
$$(\mathrm{d}/\mathrm{d}x)^k T(x) = \delta^{(k)}(x) * T(x).$$

Proof. We use induction on k. Thus assume, for a particular k, that $(\mathrm{d}/\mathrm{d}x)^k T = \delta^{(k)} * T$, and consider $k + 1$. By the previous theorem,
$$(\mathrm{d}/\mathrm{d}x)^{k+1} T = \delta' * [\delta^{(k)} * T].$$
By the associative law, this is
$$[\delta' * \delta^{(k)}] * T,$$
and applying the last theorem again we get
$$\delta^{(k+1)} * T.$$

3

Examples of distributions

In this chapter we are going to stress distributions which are not *bona fide* functions. That is, we shall emphasize those distributions which do not arise from an ordinary function $f(x)$ by the transition $f \to T_f$. (Recall that T_f is the distribution which maps each test function φ into the inner product $\langle T_f, \varphi \rangle = \int_{-\infty}^{\infty} f(x)\varphi(x) \, dx$.)

Such a distribution is the Dirac Delta Function $\delta(x)$, defined by $\langle \delta, \varphi \rangle = \varphi(0)$. As noted in Chapter 1, $\delta(x)$ cannot be a *bona fide* function. If it were, it would compress a finite amount of area into the region lying above a single point – which is impossible in integration theory.

Now, as we have seen, the derivative of any distribution is a distribution. Thus, the derivatives of the Delta Function: $\delta', \delta'', \ldots, \delta^{(n)}, \ldots$ also represent distributions, and a little thought convinces us that these distributions are even weirder than the Delta Function itself.

What good are they? Well, they have many uses. For example, the distribution $\delta^{(n)}(x)$ unifies the calculus of finite differences and the notion of an electric multipole. We shall begin by treating these two topics separately in a classical manner.

1 The finite difference calculus

Consider the finite difference operator

$$\Delta_h f(x) = f(x + h) - f(x).$$

Then the derivative is the limit

$$f'(x) = \lim_{h \to 0} h^{-1} \cdot \Delta_h f(x).$$

Iterating Δ_h we have

$$\Delta_h^2 f(x) = \Delta_h f(x+h) - \Delta_h f(x)$$
$$= f(x+2h) - f(x+h)$$
$$\underline{\qquad - f(x+h) + f(x) \qquad}$$
$$= f(x+2h) - 2f(x+h) + f(x),$$
$$\Delta_h^3 f(x) = f(x+3h) - 2f(x+2h) + f(x+h)$$
$$\underline{\qquad - f(x+2h) + 2f(x+h) - f(x) \qquad}$$
$$= f(x+3h) - 3f(x+2h) + 3f(x+h) - f(x)$$

etc. The reader undoubtedly recognizes the binomial coefficients

$$\binom{n}{k}$$

(sometimes written C_k^n),

$$\binom{n}{k} = \frac{n!}{k!\,(n-k)!}.$$

Thus we are led to the conjecture

$$\Delta_h^n f(x) = \sum_{k=0}^{n} (-1)^{n-k} \binom{n}{k} f(x+kh). \tag{1}$$

This is easily proved by induction, using the 'Pascal triangle identity'

$$\binom{n+1}{k} = \binom{n}{k-1} + \binom{n}{k}.$$

In fact the two examples above (for Δ_h^2 and Δ_h^3) show the pattern perfectly clearly, and we may dispense with any further proof.

Now a much more interesting fact is that the nth derivative is related to Δ_h^n by

$$f^{(n)}(x) = \lim_{h \to 0} h^{-n} \cdot \Delta_h^n f(x), \tag{2}$$

for C^n functions f. This can be proved using L'Hospital's rule, although such a proof (for general n) is a little tricker than might be supposed. We prefer the following conceptual proof:

$$\Delta_h f(x) = f(x+h) - f(x) = \int_0^h f'(x+u)\,du,$$

$$\Delta_h^2 f(x) = \int_0^h \frac{d}{dx} \Delta_h f(x+u)\,du$$

$$= \int_0^h \left(\frac{d}{dx} \int_0^h f'(x+u+v)\,dv \right) du$$

$$= \int_0^h \int_0^h f''(x+u+v)\,dv\,du,$$

and by induction

$$\Delta_h^n f(x) = \int_0^h \cdots \int_0^h f^{(n)}(x + u_1 + \cdots + u_n)\, du_n \cdots du_1.$$

$(n$ times$)$

Now, since $f^{(n)}$ is continuous, all of the values $f^{(n)}(x + u_1 + \cdots + u_n)$ approach $f^{(n)}(x)$ uniformly as $h \to 0$. Finally, the domain of integration is a cube of n-dimensional volume h^n. This proves (2).

(Incidentally, this proof is superior to one based on L'Hospital's rule, in that '$\int_0^h \cdots \int_0^h$ (n times)' gives the *exact* value of $\Delta_h^n f(x)$ – a fact which is sometimes useful.)

2 Dipoles, quadrupoles, 2^n-ipoles

Imagine a pair of equal and opposite electric charges $-q$ and $+q$ located a distance h apart:

$$
\begin{array}{cc}
-q & +q \\
\bullet & \bullet \\
x & x+h
\end{array}
$$

This configuration is said to have a dipole moment of qh. In particular, if we assume that $q = 1/h$, then we have a unit dipole located at (or near) the point x. The preposition 'at' (as opposed to 'near') becomes more appropriate as the distance h shrinks to zero.

Now (still setting $q = 1/h$), consider two dipoles of equal and opposite dipole moments $-q$ and $+q$ located a distance h apart. Since charges of $\pm q$ produced a unit dipole, we need charges of $\pm q^2$ in order to obtain dipole moments of size q:

$$
\begin{array}{cc}
q^2 & -q^2 \\
\bullet & \bullet \\
x & x+h
\end{array}
$$

$$
\begin{array}{cc}
-q^2 & +q^2 \\
\bullet & \bullet \\
x+h & x+2h
\end{array}
$$

which combines to give

$$
\begin{array}{ccc}
q^2 & -2q^2 & q^2 \\
\bullet & \bullet & \bullet \\
x & x+h & x+2h
\end{array}
$$

This configuration is called a unit quadrupole. Now combining two opposite quadrupoles of size q (which means that we multiply the charges again by

q, passing from $\pm q^2$ to $\pm q^3$) we have:

which combine to give

This is called a unit octopole.

By now it becomes clear that there is a strong similarity between these operations and those of the preceding section. In fact, we are led to define the unit 2^n-ipole (with a spacing of h between the charges, and with $q = 1/h$) as:

$$\text{unit } 2^n\text{-ipole} = \sum_{k=0}^{n} (-1)^{n-k} \binom{n}{k} \text{ (charges of } q^n \text{ at the point } x + kh),$$

(The 2^n, of course, refers to the fact that the sum of the binomial coefficients,

$$\sum_{k=0}^{n} \binom{n}{k} = 2^n.)$$

We can give a much more compact description of these notions if we use the Dirac Delta Function. Recall that, physically speaking, we have interpreted $\delta(t - x)$ as a unit mass located at the point $t = x$. Now if we consider electric *charge* as our paradigm, we obtain a perfectly clear interpretation of $q \cdot \delta(t - x)$ for any real constant q:

$q \cdot \delta(t - x)$ (with t as variable) = a charge q located at the point x.

Here we introduce a slightly different notation. Since we are really interested in the position x at which the charge is located, we adopt the abbreviation

$$\delta_x(t) = \delta(t - x),$$

and frequently drop the 't' and just write δ_x. It should be remembered, however, that δ_x is a generalized function (distribution) of some *other* variable t, whose action on test functions $\varphi(t)$ is given by

$$\langle \delta_x, \varphi(t) \rangle = \varphi(x).$$

Now, returning to electric multipoles (with a fixed spacing h),

$$\text{unit } 2^n\text{-ipole at } x = q^n \cdot \sum_{k=0}^{n} (-1)^{n-k} \binom{n}{k} \delta_{x+kh}$$

$$= h^{-n} \cdot \sum_{k=0}^{n} (-1)^{n-k} \binom{n}{k} \delta_{x+kh}, \qquad (3)$$

since $q = 1/h$. Now in physics the electric multipole is normally considered in situations where the spacing h is 'very small' (the physical analog of the mathematician's 'let $h \to 0$'). Thus we really want to consider

$$\lim_{h \to 0} (\text{unit } 2^n\text{-ipole at } x) = \lim_{h \to 0} h^{-n} \cdot \sum_{k=0}^{n} (-1)^{n-k} \binom{n}{k} \delta_{x+kh}.$$

What does this mean? In standard function theory it is complete nonsense. We are taking the 'limit' of a cluster of point masses whose individual size approaches infinity (at the rate h^{-n}), but which are so situated and weighted that their effects 'somehow balance out'. But, in distribution theory, all of this makes perfectly good sense. To define $\lim_{h \to 0} (\text{unit } 2^n\text{-ipole at } x)$, we simply have to find the action of it on an arbitrary test function φ; that is,

$$\lim_{h \to 0} \langle \text{unit } 2^n\text{-ipole at } x, \varphi \rangle.$$

This we have already done. By (3):

$$\lim_{h \to 0} \langle \text{unit } 2^n\text{-ipole at } x, \varphi \rangle = \lim_{h \to 0} \left\langle h^{-n} \sum_{k=0}^{n} (-1)^{n-k} \binom{n}{k} \delta_{x+kh}, \varphi \right\rangle$$

$$= \lim_{h \to 0} h^{-n} \cdot \sum_{k=0}^{n} (-1)^{n-k} \binom{n}{k} \varphi(x + kh)$$

$$= \lim_{h \to 0} h^{-n} \Delta_h^n \varphi(x)$$

$$= \varphi^{(n)}(x),$$

by (2). So the limiting value of the unit 2^n-ipole is that distribution S which maps each test function $\varphi(t)$ into $\varphi^{(n)}(x)$:

$$\text{limit of } 2^n\text{-ipole} = S, \text{ where } \langle S, \varphi \rangle = \varphi^{(n)}(x).$$

We can identify S explicitly. Since the Delta Function at x satisfies $\langle \delta_x, \varphi \rangle = \varphi(x)$, and since $\langle T', \varphi \rangle = -\langle T, \varphi' \rangle$ for any distribution T, we see that $S = (-1)^n \delta_x^{(n)}$. Thus,

$$\text{the unit } 2^n\text{-ipole at } x = (-1)^n \delta_x^{(n)},$$

or when written out in full

$$\lim_{h \to 0} h^{-n} \sum_{k=0}^{n} (-1)^{n-k} \binom{n}{k} \delta_{x+kh} = (-1)^n \delta_x^{(n)}.$$

Now the final point is this: while we have added some colorful and suggestive terminology ('multipole'), the last identity above does not depend on any of this. It is a purely mathematical theorem, which we have proved, and which could have been derived with cold-blooded rigor from the fundamental definitions in the last chapter.

3 Pseudofunctions

Since the transition from a function f to its distribution T_f is based on integration, the theory leaves out functions like $1/x^2$ which are not integrable at certain points. The treatment of such examples requires a special device, and hence these functions are frequently called 'pseudofunctions'.

Here the reader is probably expecting one of those excursions which are the hallmark of mathematical pedantry. If so, a pleasant surprise is in store. As we shall see, the treatment begins with steps which look perfectly natural, and arrives at a conclusion which seems almost paradoxical.

Our objective is to represent the function $1/x^n$ as a distribution. This example is not so special as it might seem: for many other functions with singularities can then be represented by the standard device of separating off poles. For example,

$$\frac{\cos x}{x^4} = \frac{1}{x^4} - \frac{1}{2x^2} + \frac{\cos x - 1 + (x^2/2)}{x^4},$$

the last term being continuous at $x = 0$.

The following useful lemma will be used many times throughout this book.

Lemma (The $\varphi(x)/x$ Lemma). Let $\varphi(x)$ be a C^∞ function on $\mathbb{R}^1$ such that $\varphi(0) = 0$. Then the function $\varphi(x)/x$ is C^∞.

Proof. Of course, the only difficulty occurs at $x = 0$. We use the following lovely trick. Since $\varphi(0) = 0$,

$$\frac{\varphi(x)}{x} = \frac{\varphi(x) - \varphi(0)}{x} = \int_0^1 \varphi'(xt)\, dt.$$

Then, since $\varphi \in C^\infty$, it is clear that we may differentiate under the integral sign as often as we please.   Q.E.D.

(The authors do not know who first thought of this trick, which we learned from a friend. To appreciate its beauty, try proving the above by any other method.)

Now we return to the pseudofunction $1/x^n$. We begin with $n = 1$. The pseudofunction $1/x$ is defined as the 'Cauchy principal value':

$$\left\langle \frac{1}{x}, \varphi \right\rangle = \lim_{\varepsilon \to 0} \left[\int_{-\infty}^{-\varepsilon} \frac{1}{x}\, \varphi(x)\, dx + \int_{\varepsilon}^{+\infty} \frac{1}{x}\, \varphi(x)\, dx \right].$$

The existence of this limit is a trivial consequence of the above lemma. Write $\psi(x) = [\varphi(x) - \varphi(0)]/x$, $\varphi(x) = \varphi(0) + x\psi(x)$. Now, by symmetry, the effect of the constant $\varphi(0)$ cancels out, and the 'x' in $x\psi(x)$ cancels the $1/x$ in the integral, leaving the function $\psi(x)$ which is bounded near $x = 0$.

Now, for $n > 1$, we use the fact that differentiation is always permissible in distribution theory.

Definition. We define $1/x^n$ inductively by:

$$\frac{1}{x^{n+1}} = \frac{-1}{n}\frac{d}{dx}\left(\frac{1}{x^n}\right), \quad n = \text{integer} \geqslant 1.$$

Our first result is hardly surprising.

Theorem 3.1. For $n \geqslant 0$

$$x \cdot \frac{1}{x^{n+1}} = \frac{1}{x^n}.$$

Proof. We use induction on n. The case $n = 0$ is a trivial computation. Assume now that the theorem has been proved for $n - 1$. Then,

$$\left\langle x \cdot \frac{1}{x^{n+1}}, \varphi \right\rangle = \left\langle \frac{1}{x^{n+1}}, x\varphi(x) \right\rangle$$

$$= \left\langle \frac{-1}{n}\frac{d}{dx}\frac{1}{x^n}, x\varphi(x) \right\rangle = \frac{1}{n}\left\langle \frac{1}{x^n}, \frac{d}{dx}(x\varphi(x)) \right\rangle$$

$$= \frac{1}{n}\left\langle \frac{1}{x^n}, \varphi(x) + x\varphi'(x) \right\rangle$$

$$= \frac{1}{n}\left\langle \frac{1}{x^n}, \varphi \right\rangle + \frac{1}{n}\left\langle \frac{1}{x^{n-1}}, \varphi' \right\rangle \text{ (induction hypothesis!)}$$

$$= \frac{1}{n}\left\langle \frac{1}{x^n}, \varphi \right\rangle - \frac{1}{n}\left\langle \frac{d}{dx}\left(\frac{1}{x^{n-1}}\right), \varphi \right\rangle = \frac{1}{n}\left\langle \frac{1}{x^n}, \varphi \right\rangle$$

$$+ \frac{n-1}{n}\left\langle \frac{1}{x^n}, \varphi \right\rangle. \qquad\qquad \text{Q.E.D.}$$

We define a distribution T to be *odd* or *even* if $T(-x) = -T(x)$ or $T(-x) = T(x)$, respectively. Trivially one verifies that the derivative of an odd distribution is even, and the derivative of an even distribution is odd. (Just use the identity $(d/dx)T(-x) = -T'(-x)$, proved in the last chapter.) In particular, we verify that $1/x$ is odd, and then deduce that $1/x^n$ is odd or even according to whether the integer n is odd or even, respectively.

Finally, we define a distribution T to be *positive* if

$$\langle T, \varphi \rangle \geqslant 0 \text{ for all test functions } \varphi \geqslant 0.$$

Our last result should be the statement that $1/x^n$ is positive for even exponents n. However, this is false! In fact, under suitable restrictions (see below), the distribution $1/x^{2n}$ is more nearly negative than positive. Somehow, our definitions have slipped in a negative mass of infinite magnitude situated at the origin. More precisely:

Theorem 3.2. (a) Let $\varphi(x)$ be a test function whose support does not contain the origin (i.e. φ vanishes in a neighborhood of the origin). Then

$$\left\langle \frac{1}{x^n}, \varphi \right\rangle = \int_{-\infty}^{\infty} \frac{1}{x^n} \varphi(x)\, dx.$$

(b) Let $(1/x^n)_\varepsilon$ be the function which equals $1/x^n$ for $|x| \geq \varepsilon$ and 0 elsewhere. Then the distribution $1/x^2$ is given by

$$\frac{1}{x^2} = \lim_{\varepsilon \to 0} [(1/x^2)_\varepsilon - 2\varepsilon^{-1} \cdot \delta_0].$$

Remarks. In particular, $\langle 1/x^2, \varphi \rangle$ is negative for any test function φ which takes its maximum at the origin. For the integral of $1/x^2$ over the entire domain $\{|x| \geq \varepsilon\}$ is just $2\varepsilon^{-1}$.

Part (b) could be extended to $1/x^n$, $n > 2$, except that then higher order multipoles would occur. We leave the extension as a possible exercise for the interested reader.

Proof. For part (a), we simply observe that, since we can ignore the point $x = 0$, our definition $1/x^{n+1} = (-1/n)(d/dx)(1/x^n)$ is simply a standard theorem of calculus. Thus part (a) follows from the 'Consistency Theorem', Theorem 2.2, applied to the derivative of $1/x^n$.

Part (b). By definition of the Cauchy principal value,

$$\frac{1}{x} = \lim_{\varepsilon \to 0} \left(\frac{1}{x}\right)_\varepsilon.$$

Now differentiation is a continuous operation on distribution space, and hence

$$\frac{1}{x^{n+1}} = \lim_{\varepsilon \to 0} (-1)^n \frac{1}{n!} (d/dx)^n \left(\frac{1}{x}\right)_\varepsilon.$$

Thus to find $1/x^2$ we must compute $-(d/dx)(1/x)_\varepsilon$. For $|x| > \varepsilon$ this is just $1/x^2$, and for $|x| < \varepsilon$ it is zero. However, at the points $x = \pm \varepsilon$, we get delta functions. These result from the discontinuities in the function $(1/x)_\varepsilon$, which jump by ε^{-1} at the points $x = -\varepsilon$ and $x = \varepsilon$. (Recall that, 'the derivative of a jump discontinuity is a delta function'. See the discussion of the delta function in Chapter 1.) Taking into account the minus sign in $-(d/dx)$, we have the extra term

$$-\varepsilon^{-1}[\delta_{-\varepsilon} + \delta_\varepsilon].$$

We want to replace this by

$$-2\varepsilon^{-1}\cdot\delta_0.$$

The difference is a quadrupole

$$-\varepsilon^{-1}[\delta_{-\varepsilon}-2\delta_0+\delta_\varepsilon].$$

Here the coefficient is $-\varepsilon^{-1}$. But, as we have seen, it requires coefficients of the order of ε^{-2} to give a finite quadrupole moment.

Thus, as $\varepsilon \to 0$, the difference is a quadrupole of zero moment, i.e. the zero distribution.   Q.E.D.

4

Fourier transforms

The theory of distributions achieves especial power when it is combined with the theory of Fourier transforms.

In accordance with the plan of this book, in which advanced calculus is to be the only prerequisite, we must develop the classical theory of Fourier transforms to the depth that we need it. Fortunately, this is not very far: almost all of the technical aspects of Fourier transform theory can be omitted. In fact, distribution theory provides a new approach to the whole subject – an achievement which may be the most beautiful and far reaching of all the applications of the distribution idea.

Nevertheless, there is a hard core of basic facts about the Fourier transform which we need before we can begin the distribution–theoretic treatment. Here these classical facts are laid out in a leisurely fashion, with a heavy stress on their physical motivation. The distribution–theoretic approach (which generalizes the theory to a surprising extent) is given in the next chapter.

1 The physical interpretation of complex numbers

There are many such interpretations, of course – physics is a rich subject – but the following is in some sense the clearest and most classical.

As an aside, we begin by ruling out something which is unsatisfactory. The interpretation of complex numbers as two-dimensional vectors is unsatisfying, because the curious mind asks: What about three-dimensional or n-dimensional vectors? Why such a fuss over two-dimensions? A purely mathematical answer is that the complex numbers are the only finite-dimensional extension of the reals in which all of the laws of arithmetic hold [HR, chap. 7]. These laws, of course, involve addition, subtraction, multiplication and division. Clearly multiplication and division are the kickers, and any good physical interpretation must involve multiplication as well as addition. We now give such an interpretation.

Consider a pure sinusoidal wave of fixed frequency ω. The fixed-frequency hypothesis is natural. For of all the qualities of a single coherent electro-magnetic wave (e.g. phase, amplitude, velocity, frequency), the frequency is the most permanent.

In its dependence on time t, this wave must have the form

$$f(t) = A \cdot \cos\,(2\pi\omega t - \theta),$$

where A is the *amplitude* (or maximum size) of the wave, and θ is the *phase* (or the amount by which the 'peak' of the wave is shifted away from the origin). Here θ is viewed as a fraction of a full cycle, which is 2π radians; see Figure 7.

Now, since we regard the frequency ω as fixed, the wave $f(t)$ is completely determined by its amplitude A and phase θ. We make the identification

the complex number $Ae^{i\theta}$ = a wave with amplitude A, phase θ.

At first glance this may appear a mere labelling device. It is much more, as we now show. First we recall:

The function e^{iy}. If we substitute complex values in the Taylor series $1 + z + (z^2/2!) + (z^3/3!) + \cdots$ for e^z, we arrive at the formula

$$e^{iy} = \cos y + i \sin y.$$

We *define* e^{iy} to be $\cos y + i \sin y$, and more generally:

Figure 7. A complete cycle requires an interval of length 2, so $\omega = \frac{1}{2}$. The phase $\theta = \pi/2$ or one-quarter of a cycle. That is, the peak occurs at $t = \frac{1}{2}$ which is one-quarter of the cycle length 2.

$$f(t) = A \cdot \cos\,(2\pi\omega t - \theta),$$
$$\omega = 1/2,$$
$$\theta = \pi/2.$$

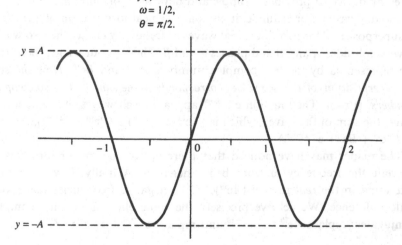

Definition. The expression e^{x+iy}, where x and y are real, is defined by

$$e^{x+iy} = e^x(\cos y + i \sin y).$$

One readily verifies that, for arbitrary complex z and w,

$$e^z \cdot e^w = e^{z+w},$$

and that for any complex constant c

$$(d/dt)e^{ct} = c \cdot e^{ct} \quad (t \text{ real}).$$

Also, $|e^{iy}| = (\cos^2 y + \sin^2 y)^{1/2} = 1$, so that $|e^{x+iy}| = e^x$.

Returning now to the wave $f(t) = A \cdot \cos(2\pi\omega t - \theta)$, we observe that $f(t)$ is the *real part* of the corresponding complex exponential form:

$$f(t) = A \cdot \cos(2\pi\omega t - \theta) = \text{Re}\,[A \cdot e^{-i(2\pi\omega t - \theta)}].$$

We take the possibly artificial step of identifying the function $f(t)$ with the complex exponential $A \cdot e^{-i(2\pi\omega t - \theta)}$, whose real part is f. The imaginary part of $A \cdot e^{-i(2\pi\omega t - \theta)}$ has no physical significance in this particular model, but it is carried along as a kind of 'shadow term', since it enables us to use the complex exponential. The merit of the complex exponential form is amply demonstrated by the following reduction formula:

$$A \cdot e^{-i(2\pi\omega t - \theta)} = (Ae^{i\theta}) \cdot e^{-2\pi i\omega t}.$$

Whether this simple formula be familiar to the reader or new, it is worth pondering. What is so special about it? Quite simply, the following.

The complex number $Ae^{i\theta}$ appears as a multiplier of the function $e^{-2\pi i\omega t}$, *which is the same for all waves of frequency* ω. Furthermore, the numbers A and θ in $Ae^{i\theta}$ represent the two standard physical parameters of the wave, namely the amplitude A and the phase θ.

Thus the identification of the complex number $Ae^{i\theta}$ with a wave of amplitude A and phase θ (the frequency ω being fixed) is far more than a labelling device. It provides a simple and natural algorithm for dealing with sinusoidal waves. For example, it gives a very neat formulation of 'principal of superposition' for two sinusoidal waves of frequency ω. Let the two waves have amplitudes A and B and phases α and β. Then, in our usage, the waves are represented by the two complex numbers $Ae^{i\alpha}$ and $Be^{i\beta}$. *Now the sum (or superposition) of the two waves corresponds to the sum of the two complex numbers.* (Proof: The function $e^{-2\pi i\omega t}$ appears in all waves of frequency ω. Thus, the sum of the waves, which is $(Ae^{i\alpha} \cdot e^{-2\pi i\omega t}) + (Be^{i\beta} \cdot e^{-2\pi i\omega t})$, reduces to $(Ae^{i\alpha} + Be^{i\beta}) \cdot e^{-2\pi i\omega t}$.)

The reader may have noticed that there is nothing in our formulas to prevent the frequency ω from being negative. Actually, if we are only interested in the *real part* of $(Ae^{i\theta})e^{-2\pi i\omega t}$, negative frequencies make very little difference. We achieve precisely the same effect if we take complex conjugates: replace $Ae^{i\theta}$ by $Ae^{-i\theta}$ and replace $e^{-2\pi i\omega t}$ by $e^{+2\pi i\omega t}$.

However, in the mathematical theory which follows, we will allow the 'frequency' ω to range over the whole real line. We will also treat the imaginary parts of functions on a par with the real parts. *The Fourier transform is a tremendously powerful mathematical tool, and it need not be tied down to any physical model.*

Now we come to the transform itself.

2 The Fourier transform

Definition (Fourier transform). Let f be a piecewise continuous integrable function on $(-\infty, \infty)$. (Readers familiar with Lebesgue theory may prefer to substitute the more general condition, $f \in L^1$.) We define the *Fourier transform* $\hat{f}$ and the *inverse Fourier transform* $\check{f}$ by the formulas

$$\hat{f}(t) = \int_{-\infty}^{\infty} e^{-2\pi itx} f(x)\, dx.$$

$$\check{f}(t) = \int_{-\infty}^{\infty} e^{2\pi itx} f(x)\, dx.$$

It will emerge presently that, under suitable conditions, $f^{\wedge\vee} = f^{\vee\wedge} = f$, so that $\check{}$ is indeed the inverse of $\hat{}$. To give a physical interpretation of $\hat{f}$, it is convenient to change the notation and write:

$$\hat{f}(t) = \int_{-\infty}^{\infty} e^{-2\pi i\omega t} f(\omega)\, d\omega.$$

Then the function $\hat{f}(t)$ appears as a (possibly quite irregular) wave, built up by superposition from the pure harmonic oscillations $e^{-2\pi i\omega t}$ of frequency ω. These oscillations are averaged together using the weight factor $f(\omega)$. The complex parameter $f(\omega)$ regulates both the amplitude and the phase with which each frequency ω enters into the representation of $\hat{f}(t)$. If the wave form $\hat{f}(t)$ is given, and we wish to determine the weight factor $f(\omega)$, we have only to compute the inverse Fourier transform $f = (\hat{f})^{\vee}$.

There are many applications of the Fourier transform which have nothing to do with 'waves'. Moreover, in case the reader has gotten tired of that subject, he may rest assured that it is in no way essential for the remainder of this book. For the most part, we shall adopt a purely mathematical viewpoint.

Other fields where the Fourier transform plays a prominent role include differential equations and probability. In each case the reason is that the Fourier transform converts some complicated operation into ordinary multiplication. (The 'complicated operations' are, respectively, differentiation and convolution. What the Fourier transform does to these is set down in the Identities Propositions below.)

Our first result about the Fourier transform is trivial but important.

Proposition. Let f be a piecewise continuous integrable function (the Lebesque theoretic assumption, $f \in L^1$, would also suffice here). Then the Fourier transform $\hat{f}$ is bounded and uniformly continuous; moreover

$$|\hat{f}(t)| \leqslant \int_{-\infty}^{\infty} |f(x)| \, dx.$$

Proof. The inequality follows immediately from the fact that $|e^{-2\pi itx}| = 1$:

$$|\hat{f}(t)| = \left| \int_{-\infty}^{\infty} e^{-2\pi itx} f(x) \, dx \right| \leqslant \int_{-\infty}^{\infty} |e^{-2\pi itx} f(x)| \, dx = \int_{-\infty}^{\infty} |f(x)| \, dx.$$

For the continuity: take $\varepsilon > 0$. Then for sufficiently large M $\int_{|x| > M} |f(x)| \, dx < \varepsilon$. But $e^{-2\pi itx}$ is uniformly continuous for $x \in [-M, M]$, whence $\hat{f}_M(t) = \int_{-M}^{M} e^{-2\pi itx} f(x) \, dx$ is uniformly continuous. Also $\hat{f}_M(t)$ differs from $\hat{f}(t)$ by less than $\int_{|x| > M} |f(x)| \, dx < \varepsilon$. Hence $\hat{f}(t)$ is the uniform limit of uniformly continuous functions.   Q.E.D.

Remarks. It is one of the worst properties of the Fourier transform that the integrability of f implies nothing about the integrability of $\hat{f}$. Thus, in general, the Fourier Inversion Theorem, which formally asserts that $f^{\wedge\sim} = f$, has no apparent meaning (because $\hat{f}$ is not integrable, and hence $f^{\wedge\sim}$ is not defined). Later, in Chapter 5, we will interpret $f^{\wedge\sim}$ in a much more general context. For the present we avoid these difficulties by imposing on our functions rather drastic 'growth conditions' (or, more properly, decay conditions) at infinity. All of the results which we obtain are of a provisional nature; they will be extended in Chapter 5. However, those extensions depend on the groundwork laid here. Detailed computation cannot be entirely avoided.

Definition. A function f on $(-\infty, \infty)$ is *rapidly decreasing* if, for every integer N, the product $x^N \cdot f(x)$ remains bounded as $x \to \pm\infty$. On the other hand, f is called *slowly increasing* if there is some integer N such that the quotient $f(x)/x^N$ remains bounded as $x \to \pm\infty$.

Observe that rapidly decreasing functions 'go down' faster than *any* power of $|x|$ as $x \to \pm\infty$, whereas slowly increasing functions may 'blow up', but their growth is dominated by *some* power of $|x|$.

Proposition. If f is rapidly decreasing and g is slowly increasing, then the product fg is rapidly decreasing. If g and h are both slowly increasing, then so is the product gh.

Proof. For fg, the 'every' in the definition of 'rapidly decreasing' kills the 'some' in the definition of 'slowly increasing'. For gh, if N_1 works for g and N_2 works for h, then $N_1 + N_2$ works for gh.

Now we introduce a class of functions which will play a key role in Chapter 5.

Definition. A function φ is called an *open support test function* (write $\varphi \in \mathscr{S}$) if φ is infinitely differentiable and every derivative $\varphi^{(k)}$ is rapidly decreasing.

(By the phrase 'every derivative' we mean $\varphi^{(k)}$ for all $k \geqslant 0$; this includes the zeroth derivative φ itself. Observe that the rate of decrease of $\varphi^{(k)}(x)$ as $|x| \to \infty$ is *not* required to be uniform in k.)

The space $\mathscr{S}$ of open support test functions is primarily important for the following reason. The Fourier transform of a function in $\mathscr{S}$ also belongs to $\mathscr{S}$. This result (the 'Closure Lemma for $\mathscr{S}$ under the Fourier transform') will be proved in due course below. But its importance should be obvious. It is nice, when dealing with an operation like the Fourier transform, to have a class of well understood functions which is closed under that operation. Thus, the C^∞ functions are important because they are closed under differentiation. The space $\mathscr{S}$ is closed under differentiation and also under the Fourier transform.

We recall that previously (in Chapter 2) we defined 'test functions' to be functions which are C^∞ with compact support. From now on, we shall refer to these as 'compact support test functions'. Thus compact support test functions *vanish identically* outside of some compact set – open support test functions are not required to vanish, but they must be rapidly decreasing together with all of their derivatives.

Examples. Every compact support test function is also an open support test function. The converse fails: thus $\varphi(x) = e^{-x^2}$ is an open support test function which does not have compact support.

Definition. A function $g(x)$ is called a *fairly good function* if g is infinitely differentiable and every derivative $g^{(k)}$ is slowly increasing.

The term 'fairly good' is due to Lighthill [**LI**]. More precisely, Lighthill calls our open support test functions 'good functions', and defines 'fairly good functions' as above.

Examples. Any polynomial is a fairly good function. But e^x is not, since it grows more rapidly than any polynomial.

The main reason for introducing fairly good functions is the following.

Proposition. The product of a fairly good function and an open support test function is an open support test function.

Proof. Let g and φ be, respectively, a fairly good function and an open support test function. We must show that $(d/dx)^k(g\varphi)$ is rapidly decreasing for every k. By Leibniz's formula

$$\left(\frac{d}{dx}\right)^k [g\varphi] = \sum_{j=0}^{k} \binom{k}{j} g^{(j)} \varphi^{(k-j)}.$$

Now, by definition, each $g^{(j)}$ is slowly increasing, and each $\varphi^{(k-j)}$ is rapidly decreasing. Thus the result follows from the fact, proved above, that (slowly increasing) × (rapidly decreasing) = (rapidly decreasing).   Q.E.D.

Convolutions. We recall the definition and some basic facts. Let $f(x)$ and $g(x)$ be piecewise continuous integrable functions. (Readers who know Lebesgue theory can substitute the more general assumption $f, g \in L^1(-\infty, \infty)$.)

Definition. We define the *convolution* $f * g$ by

$$(f * g)(x) = \int_{-\infty}^{\infty} f(x - t)g(t) \, dt.$$

Remarks. Heuristically this should be viewed as a 'sum' of translates $f_t(x) = f(x - t)$ of $f(x)$. The translate $f(x - t)$ is multiplied by the weight factor $g(t)$, and then the results are 'added' (i.e. integrated). In brief, the convolution $f * g$ is a 'weighted sum' of translates of f, and g is the weight factor.

(In some texts one finds the convolution defined as $(f * g)(x) = \int_0^x f(x - t)g(t) \, dt$. This is a special case. It corresponds to the above under the special assumption that $f(x) \equiv g(x) \equiv 0$ for $x < 0$. Thus it frequently appears in initial value problems.)

The following identities are readily verified by direct calculation:
  (i) $f * g = g * f$;
  (ii) $f * (g * h) = (f * g) * h$;
  (iii) $\dfrac{d}{dx}(f * g) = \left(\dfrac{df}{dx}\right) * g$, provided that df/dx is integrable.

Lemma (Approximate Identities Lemma). Let f, g be bounded, continuous and integrable, and suppose that $\int_{-\infty}^{\infty} g(x) \, dx = 1$. Define $g_a(x) = a \cdot g(ax)$. Then

$$(f * g_a)(x) \to f(x) \text{ as } a \to \infty.$$

The convergence is uniform in x on compact subsets of $\mathbb{R}^1$.

Proof. The idea is that, as $a \to \infty$, $g_a(x)$ becomes a narrow pulse, still having the same mass $(= 1)$ as $g(x)$, but mostly concentrated on a small interval about $x = 0$.

In more detail: take any $\delta > 0$. We break the difference $(f * g_a)(x) - f(x)$ into three parts:

$$(f * g_a)(x) - f(x) = \int_{-\infty}^{\infty} f(x - t)g_a(t)\, dt - f(x)$$

$= (A) + (B) + (C)$, where

$$(A) = \int_{|t| > \delta} f(x - t)g_a(t)\, dt,$$

$$(B) = \int_{-\delta}^{\delta} f(x - t)g_a(t)\, dt - f(x) \int_{-\delta}^{\delta} g_a(t)\, dt,$$

$$(C) = -f(x) \int_{|t| > \delta} g_a(t)\, dt.$$

Now (B) can be rewritten as

$$(B) = \int_{-\delta}^{\delta} [f(x - t) - f(x)]g_a(t)\, dt.$$

Since f is continuous and $\int_{-\infty}^{\infty} |g_a|$ is bounded independent of a, (B) $\to 0$ as $\delta \to 0$. Since f is uniformly continuous on compact subsets, the convergence of (B) to 0 is likewise uniform when x is restricted to a compact subinterval.

To make (A) and (C) small, we use the facts that g is integrable and f is bounded. Most importantly (here is where the 'narrow pulse effect' comes in),

$$\int_{|t| > \delta} |g_a(t)|\, dt = \int_{|t| > a\delta} |g(t)|\, dt,$$

and, since g is integrable,

$$\int_{|t| > a\delta} |g(t)|\, dt \to 0 \text{ as } a \to \infty.$$

Thus to make (A), (B) and (C) simultaneously 'small' take $\varepsilon > 0$; choose δ small enough to make $|(B)| < \varepsilon$; then choose a large enough to make

$$\int_{|t| > a\delta} |g(t)|\, dt = \int_{|t| > \delta} |g_a(t)|\, dt < \varepsilon.$$

Since $f(x)$ is bounded, say $|f(x)| \leqslant M$, we obtain:

$$|(B)| \leqslant \varepsilon,$$
$$|(A)| \leqslant M \cdot \varepsilon,$$
$$|(C)| \leqslant M \cdot \varepsilon,$$

and, since M is fixed, this means that the error (A) + (B) + (C) $\to 0$ as $a \to \infty$.

Q.E.D.

Further remark. There is a technical use of convolution which is so important that we must mention it here. Let f be bounded, continuous and integrable as above – but not smooth. That is, f is not assumed to be differentiable.

Let φ be an open support test function with $\int_{-\infty}^{\infty} \varphi(x)\, dx = 1$. Then, by (iii), $(f * \varphi_a) = (\varphi_a * f)$ is C^∞ for all a. By the lemma, $(f * \varphi_a) \to f$ as $a \to \infty$. Thus, the 'prickly' function f can be uniformly approximated on compact subsets by C^∞ functions. Such operators, which replace f by $\varphi_a * f$, are called 'smoothing operators' or – a name we prefer – 'mollifiers'.

3 Properties of the Fourier transform

The next proposition contains the basic facts about the Fourier transform which can be proved by direct calculation.

Proposition (Identities Proposition). Except in parts (b) and (c), the letters f, g will denote piecewise continuous integrable functions. In (b) and (c) we make the above assumptions on f, and in addition for (b) we assume that f is C^1 and f' is integrable. For (c) we assume that $x \cdot f(x)$ is integrable. The letters a and b are used to denote constants. Then,

(a) $[af(x) + bg(x)]^\wedge = a \cdot \hat{f}(t) + b \cdot \hat{g}(t)$;

(b) $[f'(x)]^\wedge = 2\pi i \cdot t \cdot \hat{f}(t)$;

(c) $[x \cdot f(x)]^\vee = (1/2\pi i)(\check{f}(t))'$;

(d) $[f(x + a)]^\wedge = e^{2\pi i a t} \cdot \hat{f}(t)$;

(e) $[e^{2\pi i a x} \cdot f(x)]^\vee = \check{f}(t + a)$;

(f) $[f(ax)]^\wedge = (1/|a|)\hat{f}(t/a)$ for $a \neq 0$;

(g) $[(f * g)(x)]^\wedge = \hat{f}(t) \cdot \hat{g}(t)$, where $*$ denotes convolution;

(h) $[f*(x)]^\wedge = \overline{\hat{f}(t)}$, where $f*(x) = \overline{f(-x)}$;

(i) $(\hat{f}, g) = (f, \hat{g})$, where (F, G) denotes the 'real inner product' $\int_{-\infty}^{\infty} F(x)G(x)\, dx$.

(Observe that we have stated (c) and (e) for the inverse transform ' $^\vee$ '. This was done to emphasize the duality between $^\wedge$ and $^\vee$. To get the ' $^\wedge$-form', we need only to recall that $\hat{f}(x) = \check{f}(-x)$.)

Proofs. Here we will do (b), (c), (g) and (i), and say something about (f). As usual, we will only sketch the main steps and leave the dreary 'ε-δ' details to the reader.

For (b): Consider $\int_{-\infty}^{\infty}$ as $\lim (M \to \infty)$ of $\int_{-M}^{M}$. Integration by parts gives

$$\int_{-M}^{M} e^{-2\pi i t x} f'(x)\, dx = -\int_{-M}^{M} (-2\pi i t)e^{-2\pi i t x} f(x)\, dx + [e^{-2\pi i t x} f(x)]_{x=-M}^{x=M}.$$

Now the fact that f and f' are *both* integrable implies that $f(x) \to 0$ as $x \to \pm\infty$. (Of course, if this were not a consequence, we could make it an assumption; it would do no harm.) Since $e^{-2\pi i t x}$ is bounded, and $f(x) \to 0$, the boundary term becomes negligible as $M \to \infty$. Thus, letting $M \to \infty$, the left side approaches $(f')^\wedge$, and the right side approaches $(2\pi i t)\hat{f}(t)$.

For (c): Differentiation under the integral sign gives, formally, $(d/dt) \int_{-\infty}^{\infty} e^{2\pi itx} f(x)\, dx = \int_{-\infty}^{\infty} (2\pi ix) e^{2\pi itx} f(x)\, dx$. To justify this we would use the fact that $\int_{-\infty}^{\infty} (1 + |x|)| f(x)|\, dx$ is finite, and estimate the difference $(e^{2\pi isx} - e^{2\pi itx})$ 'by the Mean Value Theorem'. (Strictly speaking, that theorem is not valid for complex valued functions – but, of course, one gets around this by considering the real and imaginary parts separately.)

Concerning (f): The $|a|$ in the formula for $[f(ax)]\hat{\ }$ occurs because, by definition of the Fourier transform, the integral is always taken in the positive sense, from $-\infty$ to ∞.

For (g): Interchanging the order of integration gives

$$(f * g)\hat{\ }(t) = \int_{-\infty}^{\infty} \int_{-\infty}^{\infty} e^{-2\pi itx} f(x - y) g(y)\, dy\, dx$$

$$= \int_{-\infty}^{\infty} g(y) \int_{-\infty}^{\infty} e^{-2\pi itx} f(x - y)\, dx\, dy$$

$$= \int_{-\infty}^{\infty} [e^{-2\pi ity} \hat{f}(t)] g(y)\, dy$$

(by part (d)) $= \hat{f}(t) \cdot \hat{g}(t)$.

For (i): Both inner products reduce to $\int_{-\infty}^{\infty} \int_{-\infty}^{\infty} e^{-2\pi ixy} f(x) g(y)\, dy\, dx$, except that the order of integration is reversed. The discontinuities of f and g lie along vertical and horizontal lines.

Remarks. To become fluent in the theory of Fourier transforms, it is useful to learn the above formulas by heart. Luckily, for most purposes, one doesn't have to remember the constants $(2\pi i, 1/2\pi i,$ etc.). They can be filled in 'at the last moment'. Thus $e^{2\pi iax}$ can be thought of as $e^{\text{Const} \cdot iax}$.

Whenever possible, these formulas should be visualized geometrically. For example, the operation $f(x) \mapsto f(x - a)$ represents translation by $+a$. Similarly, $f(x) \mapsto f(ax)$ (with $a > 0$) represents a stretching or shrinking in the direction of the x-axis. (Of course the stretching corresponds to small a, and the shrinking to large a.)

The special formulas $\hat{f}(0) = \int_{-\infty}^{\infty} f(x)\, dx$, and $f(0) = f\hat{\ }\check{\ }(0) = \int_{-\infty}^{\infty} \hat{f}(t)\, dt$ permit certain relationships to be seen without computation. (The latter formula involves the Fourier Inversion Theorem, to be proved below.) For example, consider the 'stretching' operation $f(x) \mapsto f(ax), 0 < a < 1$. Adopting an Alice in Wonderland viewpoint, we may think of $f(ax)$ as still being the 'same' function, with its dimensions distorted, and continue to call it f. Then $f(0)$ doesn't change (because the origin doesn't move), so $\int_{-\infty}^{\infty} \hat{f}(t)\, dt$ doesn't change. However, $\int_{-\infty}^{\infty} f(x)\, dx$ increases (the graph of f is spread out), so $\hat{f}(0)$ increases. Consequently, one is not surprised to learn that, in the horizontal direction, $\hat{f}(t)$ shrinks (the 'area' is constant), and this is precisely what the formula for $[f(ax)]\hat{\ }$ asserts.

The relations $(f')\hat{} = \text{Const} \cdot t \cdot \hat{f}(t)$ and $(x \cdot f(x))\hat{} = \text{Const} \cdot (\hat{f})'$ embody a very important principle. Roughly (in a non-technical sense), 'quickly decreasing' functions have 'smooth' Fourier transforms, and vice versa. An example is the function $e^{-|t|}$ in Figure 8b, which decreases very fast as $t \to \pm\infty$, but is not differentiable at the origin. Its inverse Fourier transform is $2/[1 + (2\pi\omega)^2]$, which is smooth but decreases rather slowly.

More technically, *rapidly decreasing* functions (in the sense of the Definition given above) have C^∞ Fourier transforms. On the other hand, if f is C^∞,

Figure 8. The Fourier transform of $f(\omega) = 2/[1 + (2\pi\omega)^2]$ (a) is $e^{-|t|}$ (b). By the Fourier inversion theorem, $\tilde{\hat{f}} = f$, this is equivalent to saying that the inverse Fourier transform of $e^{-|t|}$ is $2/[1 + (2\pi\omega)^2]$. The inverse transform is easily computed by elementary calculus: $\int_{-\infty}^{\infty} e^{2\pi i\omega t} e^{-|t|} dt = 2 \cdot \int_0^\infty \cos 2\pi\omega t \cdot e^{-t} dt = 2/[1 + (2\pi\omega)^2]$. (The direct transform of $2/[1 + (2\pi\omega)^2]$ can only be computed by contour integrals or equivalent tools. This translation of an elementary integral – above – into a non-elementary one, is typical of the Fourier transform.) Now we apply the standard physical interpretation to this particular case. The irregular wave whose form is $e^{-|t|}$ is built up from a continuum of 'pure colors' $e^{-2\pi i\omega t}$, the different frequencies ω being weighted in proportion to $2/[1 + (2\pi\omega)^2]$.

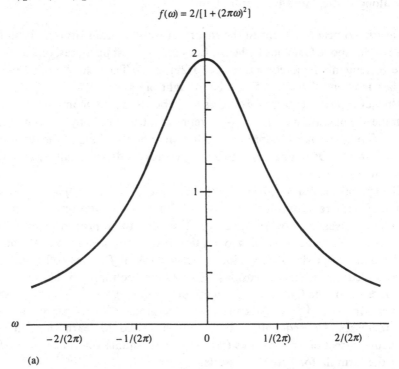

$$f(\omega) = 2/[1 + (2\pi\omega)^2]$$

(a)

and each derivative $f^{(k)}$ is integrable, then $\hat{f}$ is rapidly decreasing. The proofs are obvious, after Identities (b) and (c). Unfortunately neither of these statements is the converse of the other. The next result is more symmetrical. As mentioned earlier, this result is the principal reason for the singular importance of the space $\mathscr{S}$.

Lemma (Closure of $\mathscr{S}$ under the Fourier transform). The Fourier transform of an open support test function is an open support test function.

Proof. Let φ be an open support test function. We need to show that $\hat{\varphi}^{(k)}$ is rapidly decreasing, i.e. that $t^N \cdot \hat{\varphi}^{(k)}(t)$ is bounded as a function of t for every fixed N and k. Now $t^N \cdot \hat{\varphi}^{(k)}(t)$ is the Fourier transform of $\text{Const} \cdot (d/dx)^N(x^k \cdot \varphi(x))$. (The constant can be worked out, of course, but its value doesn't concern us.) From the Leibniz formula for derivatives:

$$\left(\frac{d}{dx}\right)^N (x^k \cdot \varphi(x)) = \sum_{i=0}^{N} \text{Const}_i \cdot x^{k-i} \cdot \varphi^{(N-i)}(x).$$

(Again the constants don't matter; they depend on the integers N, k, i but not on the function φ.) We observe that $\text{Const}_i = 0$ for $i > k$. Now each of the functions $x^{k-i} \cdot \varphi^{(N-i)}(x)$ is rapidly decreasing. And rapidly decreasing functions are clearly integrable, by comparison, e.g., with $1/(1 + x^2)$. Finally,

Figure 8 (continued)

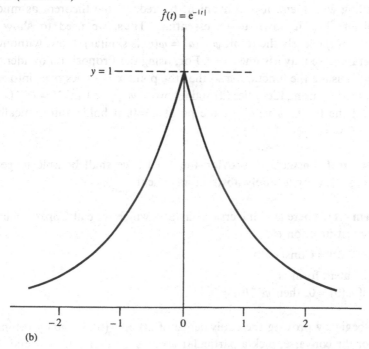

$\hat{f}(t) = e^{-|t|}$

$y = 1$

$-2 \qquad -1 \qquad 0 \qquad 1 \qquad 2$

(b)

integrable functions have bounded Fourier transforms. So $t^N \hat{\phi}^{(k)}(t)$ is bounded, as desired.   Q.E.D.

The following is the only result in this chapter which we dignify with the title 'Theorem'.

Theorem 4.1 (Fourier Inversion Theorem). For any test function φ, $\varphi^{\hat{}\check{}} = \varphi^{\check{}\hat{}} = \varphi$.

Corollary 4.1a. The Fourier transform, viewed as a mapping from $\mathscr{S}$ to $\mathscr{S}$, is one to one and onto ($\mathscr{S}$ = space of test functions).

Corollary 4.1b. If $\varphi, \psi \in \mathscr{S}$, then $(\varphi\psi)^{\hat{}} = \hat{\varphi} * \hat{\psi}$.

Proof of corollaries (assuming the theorem). For Corollary 4.1a: the preceding lemma shows that $\hat{}$ maps $\mathscr{S}$ into $\mathscr{S}$; $\check{\phi}(t)$ is just $\hat{\phi}(-t)$, so the theorem implies that $\hat{}$ has an inverse which maps $\mathscr{S}$ into $\mathscr{S}$. This proves the first corollary.

For Corollary 4.1b: by Identity (g) we have $(F * G)^{\check{}} = \check{F}\check{G}$ (it works for $\check{}$ as well as $\hat{}$), so $(\hat{\varphi} * \hat{\psi})^{\check{}} = (\varphi^{\hat{}\check{}})(\psi^{\hat{}\check{}}) = \varphi\psi$ (by the Inversion Theorem) and applying the theorem again gives $\hat{\varphi} * \hat{\psi} = (\varphi\psi)^{\hat{}}$.

We will give two proofs of the Fourier Inversion Theorem (F.I.T). Before proceeding with them, it is convenient to 'reduce' the theorem as much as possible, boiling it down to its essentials. Thus, we need to show that $\varphi^{\hat{}\check{}}(a) = \varphi(a)$ (clearly the result $\varphi^{\check{}\hat{}}(a) = \varphi(a)$ is similar). Now, without loss of generality, we may assume $a = 0$. For, using the Proposition on identities, we can translate the function φ so that the point $x = a$ goes over into $x = 0$. More precisely, using Identities (d) and (e), we have $[\varphi(x + a)]^{\hat{}\check{}} = \varphi^{\hat{}\check{}}(x + a)$. Hence, if the F.I.T. is valid for the point $x = 0$, it holds automatically for all points $x = a$.

The universal constant. In proving the F.I.T., we shall be able to get the following result by a purely abstract argument.

Theorem 4.1*. There is a universal constant, which we call Const, such that, for any test function φ,

    (a) $\varphi^{\hat{}\check{}}(0) = \text{Const} \cdot \varphi(0)$.

An equivalent form is

    (b) if $\varphi(0) = 0$, then $\varphi^{\hat{}\check{}}(0) = 0$.

We begin by proving the equivalence of (a) and (b). Trivially (a) implies (b). For the converse, pick a particular $\psi \in \mathscr{S}$ such that $\psi(0) = 1$, and define

Const $= \psi^{\wedge\vee}(0)$. Then, given any test function φ, applying (b) to the difference $\varphi - \varphi(0) \cdot \psi$ yields the result (a).

Next we show that Theorem 4.1* implies Theorem 4.1. This amounts to showing that the constant in Theorem 4.1* is equal to 1.

To prove that Const $= 1$, we must find a particular function φ for which $\varphi^{\wedge\vee}(0)$ can be computed. One possibility would be $\varphi(x) = e^{-|x|}$ but this is not a test function. We shall use the function $e^{-\pi x^2}$.

Consider the integral $I = \int_{-\infty}^{\infty} e^{-x^2}\, dx$. Then $I^2 = \int_{-\infty}^{\infty} \int_{-\infty}^{\infty} e^{-x^2} \cdot e^{-y^2}\, dy\, dx$. Putting this in polar coordinates gives $I^2 = \int_0^{2\pi} \int_0^{\infty} e^{-r^2} r\, dr\, d\theta = \pi$, so $I = \sqrt{\pi}$. By an elementary change of variables, $\int_{-\infty}^{\infty} e^{-\pi x^2}\, dx = 1$.

Now let $\varphi(x) = e^{-\pi x^2}$. Then φ satisfies the differential equation $\varphi'(x) = (-2\pi x) \cdot \varphi(x)$, and taking the Fourier transform of this equation, using Identities (b) and (c), gives $(\hat{\varphi})' = (-2\pi t)\hat{\varphi}$: the same differential equation! This equation has a unique solution, up to a constant, so $\hat{\varphi}(x) = \text{Konst} \cdot e^{-\pi x^2}$. Furthermore, Konst $= \hat{\varphi}(0) = \int_{-\infty}^{\infty} e^{-\pi x^2}\, dx = 1$. Thus $\hat{\varphi} = \varphi$, and hence clearly $\varphi^{\wedge\vee} = \varphi$ (for this particular function φ). This proves that the universal constant is 1.

Now all that remains is to prove Theorem 4.1*. Of course, that is the essential step. Indeed, everything we have done so far has involved rather trivial manipulations. The essence of the Fourier Inversion Theorem is contained in Theorem 4.1*. We give two proofs of it.

First proof of Theorem 4.1* (following Hörmander [HO]). We shall prove the form (a), that there exists a universal constant 'Const' such that, for all test functions φ, $\varphi^{\wedge\vee}(0) = \text{Const} \cdot \varphi(0)$. This proof is valid under the more general assumptions that φ is continuous, bounded and integrable, and $\hat{\varphi}$ is integrable.

The proof is based on the Approximate Identities Lemma for convolutions, given earlier in this chapter.

Take any particular test function ψ such that $\psi(0) = 1$, $\psi(x) \geqslant 0$, and $\hat{\psi}(t) \geqslant 0$. (Such functions ψ do exist, e.g. $\psi(x) = e^{-\pi x^2}$.) Let Const $= \int_{-\infty}^{\infty} \hat{\psi}(t)\, dt$. The Const > 0 (since $\hat{\psi} \geqslant 0$ and $\hat{\psi}(0) \neq 0$). This, as we shall show, is the desired universal constant.

For $a > 0$, write $\psi_a(x) = \psi(x/a)$, so that, by Identity (f), $\hat{\psi}_a(t) = a\hat{\psi}(at)$. By Identity (i), the inner product

$$(\varphi, \hat{\psi}_a) = (\hat{\varphi}, \psi_a).$$

Now the idea of the proof is that, as $a \to \infty$, $\hat{\psi}_a$ becomes a narrow pulse – an 'approximate identity' in the sense of the lemma – whereas ψ_a spreads out and becomes like the constant function 1. This implies as we shall show, that as $a \to \infty$

$$(\varphi, \hat{\psi}_a) \to \text{Const} \cdot \varphi(0),$$

$$(\hat{\varphi}, \psi_a) \to \int_{-\infty}^{\infty} \hat{\varphi}(t)\, dt.$$

These identities, combined with $(\varphi, \hat{\psi}_a) = (\hat{\varphi}, \psi_a)$, will give the desired result. Here are the details. Define

$$g(t) = (1/\text{Const}) \cdot \hat{\psi}(-t)$$
$$g_a(t) = (1/\text{Const}) \cdot \hat{\psi}_a(-t) = (1/\text{Const})a\hat{\psi}(-at),$$

so that $g_a(t) = ag(at)$ as in the Approximate Identities Lemma. Note also that, by definition of Const, $\int_{-\infty}^{\infty} g(t)\,dt = 1$. Hence, by the lemma,

$$(g_a * \varphi)(x) = (1/\text{Const}) \int_{-\infty}^{\infty} \hat{\psi}_a(t - x)\varphi(t)\,dt \to \varphi(x) \text{ as } a \to \infty.$$

Setting $x = 0$ we have

$$\int_{-\infty}^{\infty} \varphi(t)\hat{\psi}_a(t)\,dt = (\varphi, \hat{\psi}_a) \to \text{Const} \cdot \varphi(0) \text{ as } a \to \infty.$$

On the other hand, as $a \to \infty$, $\psi_a(x) = \psi(x/a) \to \psi(0) = 1$ uniformly on compact subsets. Hence, because $\hat{\varphi}$ is integrable,

$$(\hat{\varphi}, \psi_a) \to \int_{-\infty}^{\infty} \hat{\varphi}(t)\,dt \text{ as } a \to \infty.$$

Thus, we have established the identities for $(\varphi, \hat{\psi}_a)$ and $(\hat{\varphi}, \psi_a)$ promised above. This proves the theorem.

Query. We proved the F.I.T. by showing that $\varphi(0) = \int_{-\infty}^{\infty} \hat{\varphi}(t)\,dt$. Now since $\check{\varphi}(t) = \hat{\varphi}(-t)$, it is equally true that $\varphi(0) = \int_{-\infty}^{\infty} \check{\varphi}(t)\,dt$. What in our proof implies that $\check{}$, is the inverse transform of $\hat{}$? (See the 'without loss of generality' step above, where we showed that we could take $a = 0$.)

Our second proof of the F.I.T. involves the '$\varphi(x)/x$ lemma' introduced in Chapter 3. To maintain flow, and because this lemma is so important, we shall repeat it here.

The $\varphi(x)/x$ Lemma. If $\varphi \in \mathscr{S}$ and $\varphi(0) = 0$, then $\varphi(x)/x$ belongs to $\mathscr{S}$.

Proof. The only difficulty is to show that $\varphi(x)/x$ is C^∞ at $x = 0$. As in Chapter 3: since $\varphi(0) = 0$, $\varphi(x)/x = \int_0^1 \varphi'(xt)\,dt$. Thus, differentiation under the integral sign yields the desired result.   Q.E.D.

The second proof of the Inversion Theorem is less natural than the first one. However, it is very short.

Second proof of Theorem 4.1*. We prove the equivalent form (b), namely that $\varphi(0) = 0$ implies $\int_{-\infty}^{\infty} \phi(t)\, dt = 0$. Take any test function φ with $\varphi(0) = 0$. Set $\psi(x) = \varphi(x)/x$. Then $\psi \in \mathscr{S}$ by the above lemma. Hence $\hat{\psi} \in \mathscr{S}$, $\hat{\psi}$ is rapidly decreasing, and $\hat{\psi}(\pm\infty) = 0$. Since

$$\varphi(x) = x\psi(x), \ \phi(t) = \check{\phi}(-t) = (-1/2\pi i)(\hat{\psi}(t))' \ (\text{Identity (c)}).$$

Thus, the Fundamental Theorem of Calculus gives:

$$\int_{-\infty}^{\infty} \phi(t)\, dt = (-1/2\pi i) \int_{-\infty}^{\infty} (\hat{\psi}(t))'\, dt = (-1/2\pi i)[\hat{\psi}(\infty) - \hat{\psi}(-\infty)] = 0.$$

Q.E.D.

5

Tempered distributions

The tempered distributions are a subset of the set of all distributions. For clarity, we shall refer to the latter as *general distributions*. The motive for introducing tempered distributions is that they behave particularly well with respect to the Fourier transform.

As was already suggested in the preceding chapter, the Fourier Inversion Theorem is the foundation of Fourier transform theory. Obviously one wants to have this theorem in as general a form as possible. Quite simply, this allows calculations to be performed without a continual 'stop and think' hesitation each time the Fourier Inversion Theorem is invoked. It is our considered judgment that no previous theory has achieved the degree of elegance and generality that the theory of tempered distributions has. The problem was not that the facts were unknown – they have been known for a long time – but there was always a certain awkwardness in their presentation. To show the nature of the difficulties, we shall glance briefly at one of the earlier presentations – that based on L^1 (integrable) functions. Since the Fourier transform is given by an integral, $\hat{f}(t) = \int_{-\infty}^{\infty} e^{-2\pi itx} f(x)\, dx$, it is natural to require that this integral converge, and this in turn requires the integrability of the function $f(x)$. Where is the trouble? Well, the Fourier transform $\hat{f}(t)$ is not necessarily integrable. Hence the Fourier Inversion Theorem (which asserts that $f(x)$ is the inverse Fourier transform of $\hat{f}(t)$) becomes meaningless. Actually, it can be given a meaning, but special – and rather *ad hoc* – methods are necessary to salvage it. With the theory of tempered distributions, all of these difficulties disappear.

There are two differences between tempered and general distributions, one very minor and the other serious. The minor distinction is that, whereas we considered general distributions in both the real and complex case (with real/complex valued test functions, scalars, etc.) tempered distributions are only interesting in the complex case. That is because we wish to apply the Fourier transform – a complex operation.

The major quality which distinguishes tempered distributions from general distributions is that the former act on a different class of test functions. Namely, tempered distributions act on the open support test functions which were introduced in the preceding chapter. Let us reconnoiter. A general distribution is a continuous mapping from the set of (compact support) test functions into the real or complex numbers. A tempered distribution is a continuous mapping from the open support test functions into the complex numbers. The compact support test functions are a proper subset of the class of open support test functions. For clearly, if a function has compact support, then it is rapidly decreasing together with all of its derivatives – since it vanishes identically near infinity. In a parallel fashion, we shall see that the topology on the compact support test functions is stronger (more stringent) than the topology on the open support test functions. This has the following logical consequences. Any tempered distribution (which acts on the larger class of open support test functions) is automatically a general distribution (acting on the smaller set of compact support test functions). As noted above, this is not only true set theoretically, but also topologically. Since the topology on the open support test functions is weaker, a linear functional continuous in terms of this weaker topology is also continuous in terms of the stronger compact support topology. Let us spell this out.

In the preceding paragraph, we have used the term 'topology' rather glibly, since our presentation in this book is based on the convergence of sequences. More precisely, then, we have the following. When we say that the topology on the compact support test functions is stronger, we mean that if $\varphi_n \to \varphi$ in the compact support sense ($\varphi_n \in \mathcal{D}$), then $\varphi_n \to \varphi$ in the open support sense. Now the rest is pure logic. Continuity (in either sense) means that if $\varphi_n \to \varphi$ (in the sense considered), then $\langle T, \varphi_n \rangle \to \langle T, \varphi \rangle$. Here the conclusion $\langle T, \varphi_n \rangle \to \langle T, \varphi \rangle$ involves only the ordinary convergence of complex numbers. The condition $\varphi_n \to \varphi$ lies in the 'if' part of the above 'if–then' statement. Hence, the stronger the topology on the φ_n, the easier continuity is to attain, and the more continuous functionals there will be.

For convenience, we restate one definition from Chapter 4.

Definition. An *open support test function* $\varphi(x)$ is a complex valued function of the real variable x such that
  (a) $\varphi(x)$ is C^∞,
  (b) every derivative $\varphi^{(k)}(x)$ is rapidly decreasing (i.e., for every integer N, the product $x^N \cdot \varphi^{(k)}(x)$ remains bounded as $x \to \pm\infty$).

Closely related to the above is the following:

Definition. A sequence of open support test functions $\varphi_n(x)$ *converges to zero* (write $\varphi_n \to 0$) if, for each pair of integers N and k, the sequence of functions $x^N \cdot \varphi_n^{(k)}(x)$ approaches zero uniformly as $n \to \infty$.

Of course, we say that $\varphi_n \to \varphi$ if $(\varphi_n - \varphi) \to 0$.

An alternative form of this definition, which is sometimes useful, is the following. Its merit is that, for all x, the multiplier $(1 + |x|)^N$ increases as N increases. (By contrast, $|x|^N$ increases with N only for $|x| > 1$.)

Equivalent definition. $\varphi_n \to 0$ if, for each pair of integers N and k, the sequence

$$(1 + |x|)^N \varphi_n^{(k)}(x) \to 0$$

uniformly as $n \to \infty$.

The set of all open support test functions is denoted by $\mathscr{S}$. We will use the term 'convergence in $\mathscr{S}$' for the convergence defined above.

Recall that a function $f(x)$ is *rapidly decreasing* if, for each N, $x^N \cdot f(x) \to 0$ as $|x| \to \infty$; and a function $g(x)$ is *slowly increasing* if there exists some integer N such that $g(x)/x^N \to 0$ as $|x| \to \infty$.

From Chapter 4 we recall the obvious but very important fact: *the product $g(x)f(x)$ of a slowly increasing function g and a rapidly decreasing function f is rapidly decreasing.*

We also recall from Chapter 4 the:

Proposition. Let $g(x)$ be a C^∞ function which is slowly increasing together with all of its derivatives, and let $\varphi(x)$ be an open support test function. Then $g(x)\varphi(x)$ is an open support test function.

Remark. It will follow below that a function $g(x)$ as above can be multiplied by any tempered distribution. It can be proved that these are the only functions with this property. Thus we get a very lopsided theory of multiplication: allowing complete freedom for the distribution T but imposing severe restrictions on the function $g(x)$. This is one of the worst aspects of the theory. It will be remedied in Chapter 7.

1 Basic definitions and facts

This section follows very closely the plan laid out in Chapter 2. Actually, since the tempered distributions are a *subset* of the set of general distributions, certain parts of this chapter are corollaries of the results in Chapter 2. This is true, for example, of the calculus identities – since they hold for the wider class of general distributions, they automatically retain their validity for tempered distributions. On the other hand, where topological considerations are involved, we have to be a little more careful.

The following proposition disposes of one of these topological points.

Proposition. The set $\mathscr{D}$ of compact support test functions is dense in the set $\mathscr{S}$ of open support test functions, in terms of convergence in $\mathscr{S}$. Hence, every tempered distribution (i.e. every element of $\mathscr{S}'$) corresponds to a unique distribution in $\mathscr{D}'$.

Proof. We use 'mesa functions' as introduced in Chapter 1. Recall that such a function $\psi(x)$ is C^∞ with compact support, $0 \leqslant \psi(x) \leqslant 1$, and $\psi(x) \equiv 1$ on some neighborhood of $x = 0$. Let $\psi_a(x) = \psi(x/a)$. Then ψ_a has compact support, so that $\psi_a \in \mathscr{D}$. For any function $\varphi \in \mathscr{S}$ ($\mathscr{S}$ not $\mathscr{D}$), the product $\psi_a \varphi \in \mathscr{D}$. Hence it suffices to prove the following lemma, which will be used several times throughout the remainder of this book.

Mesa Function Lemma. Let ψ be a mesa function, as defined in the preceding paragraph. Let $\psi_a(x) = \psi(x/a)$. Take any function $\varphi \in \mathscr{S}$. Then, as $a \to \infty$, the product $\psi_a \varphi \to \varphi$ in $\mathscr{S}$.

Proof of lemma. The idea is that $\psi_a(x)$ 'spreads out towards infinity' as $a \to \infty$. Thus $\psi_a(x) \to 1$ as $a \to \infty$, uniformly on any compact interval $-M \leqslant x \leqslant M$. Also, $\psi_a(x)$ remains bounded (between 0 and 1) for all a and x. Lastly, and most importantly, the derivatives $\psi_a^{(k)}(x) \to 0$ uniformly as $a \to \infty$, for each fixed $k \geqslant 1$. Now take any $\varphi \in \mathscr{S}$, and apply the Leibniz rule:

$$\left(\frac{\mathrm{d}}{\mathrm{d}x}\right)^k [\psi_a(x)\varphi(x)] = \sum_{j=0}^{k} \binom{k}{j} \psi_a^{(j)}(x)\varphi^{(k-j)}(x).$$

From the above stated facts about ψ_a and the fact that $\varphi \in \mathscr{S}$, we immediately deduce:

$$(1 + |x|)^N \left(\frac{\mathrm{d}}{\mathrm{d}x}\right)^k [\psi_a(x)\varphi(x)] \to (1 + |x|)^N \left(\frac{\mathrm{d}}{\mathrm{d}x}\right)^k \varphi(x)$$

as $a \to \infty$, fo each fixed k and N, and uniformly in x. This is one of the definitions of convergence in $\mathscr{S}$. Hence, as $a \to \infty$, the functions $\psi_a \varphi$, with compact support, converge in $\mathscr{S}$ to the function φ.

Now we come to the main definition.

Definition. A *tempered distribution* T is a mapping from the set of open support test functions into the complex numbers which satisfies the following:
  (a) (Linearity.) $\langle T, a\varphi(x) + b\psi(x) \rangle = a \cdot \langle T, \varphi(x) \rangle + b \cdot \langle T, \psi(x) \rangle$ for all open support test functions φ, ψ and all complex constants a, b;
  (b) (Continuity.) If $\varphi_n \to 0$ in $\mathscr{S}$, then $\langle T, \varphi_n \rangle \to 0$.
The set of all tempered distributions is denoted by $\mathscr{S}'$.

Definition. Let $f(x)$ be a slowly increasing piecewise continuous function. Then we define the distribution T_f corresponding to f by

$$\langle T_f, \varphi \rangle = \int_{-\infty}^{\infty} f(x)\varphi(x)\,\mathrm{d}x$$

for all open support test functions φ.

The next definition is repeated practically verbatim from Chapter 2. We restate it here for convenience. Also we note that the Continuity Theorem 5.1 below is *not* a consequence of the corresponding theorem in Chapter 2, because the topology on $\mathscr{S}$ differs from that on $\mathscr{D}$. This point can be clarified in the following way. We already know that every tempered distribution corresponds to a unique general distribution (although not conversely), and that calculus operations like $\mathrm{d}/\mathrm{d}x$ act on general distributions. What we need to show is that the derivative of a tempered distribution is again a tempered distribution, and this is a consequence of Theorem 5.1.

Definition (calculus operations). Let S and T be tempered distributions. Then we define new distributions $S + T$, cT (for complex constants c), $(\mathrm{d}/\mathrm{d}x)T(x)$, $T(ax)$ and $T(x - a)$ (for real constants $a \neq 0$), and $g(x)T(x)$ (where $g(x)$ is a C^∞ function which is slowly increasing together with all of its derivatives) by

(1) $\langle S + T, \varphi \rangle = \langle S, \varphi \rangle + \langle T, \varphi \rangle$;

(2) $\langle cT, \varphi \rangle = c\langle T, \varphi \rangle$;

(3) $\langle (\mathrm{d}/\mathrm{d}x)T(x), \varphi \rangle = -\langle T, (\mathrm{d}/\mathrm{d}x)\varphi(x) \rangle$;

(4) $\langle T(ax), \varphi \rangle = |a|^{-1}\langle T, \varphi(x/a) \rangle$;

(5) $\langle T(x - a), \varphi \rangle = \langle T, \varphi(x + a) \rangle$;

(6) $\langle g(x)T(x), \varphi \rangle = \langle T, g(x)\varphi(x) \rangle$.

The following definition is new – it has no analog in Chapter 2.

Definition (Fourier transform). The Fourier transform $\hat{T}$ and inverse Fourier transform $\check{T}$ of a tempered distribution T are defined by:

(7) $\langle \hat{T}, \varphi \rangle = \langle T, \hat{\varphi} \rangle$;

(7a) $\langle \check{T}, \varphi \rangle = \langle T, \check{\varphi} \rangle$.

We note that $\check{T}(x) = \hat{T}(-x)$ because $\check{\varphi}(x) = \hat{\varphi}(-x)$.

Theorem 5.1 (Continuity Theorem). All of the operations in the preceding definitions are tempered distributions.

Proof. Since condition (a) (linearity) is trivial, we turn to condition (b) (continuity). We give proofs for T_f and $\hat{T}$. The proof for T_f shows how we handle the circumstances that the test functions $\varphi \in \mathscr{S}$ lack compact support.

After this has been done here (once!) we need not repeat all of the similar arguments in the theorems which follow.

Consider T_f and let $\varphi_n \to 0$ in $\mathscr{S}$. We need to show that $\langle T_f, \varphi_n \rangle = \int_{-\infty}^{\infty} f(x)\varphi_n(x)\,dx \to 0$. Now since f is slowly increasing, there is an integer N and a constant such that $|f(x)| < \text{Const} \cdot (1 + |x|)^N$. On the other hand, since $\varphi_n \to 0$ in $\mathscr{S}$, there is a sequence of positive numbers ε_n converging to zero such that $|\varphi_n(x) \cdot (1 + |x|)^{N+2}| \leqslant \varepsilon_n$, i.e. $|\varphi_n(x)| \leqslant \varepsilon_n \cdot (1 + |x|)^{-N-2}$. Hence,

$$|\langle T_f, \varphi_n \rangle| = \left| \int_{-\infty}^{\infty} f(x)\varphi_n(x)\,dx \right| \leqslant \varepsilon_n \cdot \text{Const} \cdot \int_{-\infty}^{\infty} (1 + |x|)^{-2}\,dx \to 0 \text{ as } n \to \infty$$

because $\int_{-\infty}^{\infty} (1 + |x|)^{-2}\,dx$ is finite. Thus T_f is a tempered distribution.

We now turn to the continuity of $\hat{T}$. We need the following:

Lemma. The Fourier transform is continuous as a mapping from $\mathscr{S}$ to $\mathscr{S}$. That is, if $\varphi_n \to 0$ in $\mathscr{S}$, then $\hat{\varphi}_n \to 0$ in $\mathscr{S}$.

Proof. The proof is similar to that of the Closure Lemma in Chapter 4, which showed that if $\varphi \in \mathscr{S}$, then $\hat{\varphi} \in \mathscr{S}$.

To show that $\hat{\varphi}_n \to 0$ in $\mathscr{S}$, we need to show that

$$t^N(\hat{\varphi}_n)^{(k)}(t) \to 0 \text{ uniformly for each } N \text{ and } k.$$

Now, from the Identities Proposition in Chapter 4,

$$t^N(\hat{\varphi}_n)^{(k)}(t) = \text{Const} \cdot [(x^k\varphi_n(x))^{(N)}]\hat{\ }$$

From the definition of convergence in $\mathscr{S}$, $\varphi_n \to 0$ in $\mathscr{S}$ implies $[x^k \cdot \varphi_n(x)]^{(N)} \to 0$ in $\mathscr{S}$. Hence it suffices to prove that

$$\theta_n \to 0 \text{ in } \mathscr{S} \text{ implies } \hat{\theta}_n(t) \to 0 \text{ uniformly.}$$

(Then, of course, we set $\theta_n(x) = [x^k \cdot \varphi_n(x)]^{(N)}$.)

Recall that for Fourier transforms in general we have, since $|e^{-2\pi itx}| = 1$,

$$|\hat{f}(t)| = \left| \int_{-\infty}^{\infty} e^{-2\pi itx} f(x)\,dx \right| < \int_{-\infty}^{\infty} |f(x)|\,dx.$$

Thus to make $\hat{\theta}_n(t) \to 0$ uniformly, we have only to ensure that

$$\int_{-\infty}^{\infty} |\theta_n(x)|\,dx \to 0.$$

This follows easily from the fact that $\theta_n \to 0$ in $\mathscr{S}$. Indeed, $\theta_n \to 0$ in $\mathscr{S}$ implies in particular that $(1 + |x|)^2\theta_n(x) \to 0$ uniformly. Since $1/(1 + |x|)^2$ is integrable over $-\infty < x < \infty$, we obtain

$$\int_{-\infty}^{\infty} |\theta_n(x)|\,dx = \int_{-\infty}^{\infty} (1 + |x|)^2|\theta_n(x)| \frac{1}{(1 + |x|)^2}\,dx \to 0.$$

This proves the lemma.

Now we return to the continuity of $\hat{T}$. As usual, once we have the necessary results for test functions, the rest is easy.

Continuity for $\hat{T}$ means that, if $\varphi_n \to 0$ in $\mathscr{S}$, then $\langle \hat{T}, \varphi_n \rangle \to 0$. By definition of $\hat{T}$, $\langle \hat{T}, \varphi_n \rangle = \langle T, \hat{\varphi}_n \rangle$. We have just proved that $\varphi_n \to 0$ in $\mathscr{S}$ implies $\hat{\varphi}_n \to 0$ in $\mathscr{S}$. Now T itself – being a distribution – is continuous by definition. Thus $\hat{\varphi}_n \to 0$ in $\mathscr{S}$ implies $\langle T, \hat{\varphi}_n \rangle \to 0$. Putting all of this together: $\varphi_n \to 0$ in $\mathscr{S}$ implies $\hat{\varphi}_n \to 0$ in $\mathscr{S}$ implies $\langle T, \hat{\varphi}_n \rangle = \langle \hat{T}, \varphi_n \rangle \to 0$. This proves the continuity of $\hat{T}$.

Now we define convergence for a sequence of tempered distributions. Just as convergence of test functions gives a topology to $\mathscr{S}$, the definition below imposes a topology on $\mathscr{S}'$.

Definition. Let $\{T_n\}$ be a sequence of tempered distributions, and let T be a tempered distribution. We say that T_n *converges to* T, written $T_n \to \mathrm{T}$, if

$$\langle T_n, \varphi \rangle \to \langle T, \varphi \rangle$$

for every open support test function φ.

The next three theorems are virtual corollaries of the corresponding theorems for general distributions in Chapter 2. We shall compress the statements of these theorems except for the 'calculus identities', which, since they form the basis for computations, are restated in full.

Theorem 5.2 (Consistency Theorem). (Compare Theorem 2.2.) Let f be a piecewise C^1 function which is slowly increasing together with its derivative. Then

$$(T_f)' = T_{f'(x)}.$$

Similar statements hold for the other operations (addition, scalar multiplication, etc.) which we have defined for tempered distributions. In particular, if f is an integrable function and $\hat{f}(t)$ is its Fourier transform (in the classical sense), then

$$(T_f)\hat{} = T_{\hat{f}}$$

Proof. We put aside for a moment a discussion of the Fourier transform. Otherwise, since each tempered distribution corresponds to a unique general distribution, these results are corollaries of the earlier Consistency Theorem 2.2.

The Fourier transform must be treated in detail, since it was not discussed in Chapter 2. Namely, we must show that our definition $\langle \hat{T}, \varphi \rangle = \langle T, \hat{\varphi} \rangle$ is the 'correct' one: as explained in Chapters 1 and 2, 'correctness' means that the appropriate consistency theorem holds.

Let $f(x)$ be any integrable function, and let $\hat{f}(t)$ be its Fourier transform. Let $\varphi \in \mathscr{S}$ be any open support test function. Then, by the classical definition of the Fourier transform,

$$\hat{f}(t) = \int_{-\infty}^{\infty} e^{-2\pi i t x} f(x) \, dx.$$

Hence,

$$\langle T_{\hat{f}(t)}, \varphi(t) \rangle = \int_{-\infty}^{\infty} \int_{-\infty}^{\infty} e^{-2\pi i t x} f(x) \varphi(t) \, dx \, dt.$$

Now we consider the distribution definition of the Fourier transform:

$$\langle \hat{T}_{f(x)}, \varphi(x) \rangle = \langle T_{f(x)}, \hat{\varphi}(x) \rangle = \left\langle T_{f(x)}, \int_{-\infty}^{\infty} e^{-2\pi i t x} \varphi(t) \, dt \right\rangle$$

$$= \int_{-\infty}^{\infty} \int_{-\infty}^{\infty} e^{-2\pi i t x} f(x) \varphi(t) \, dt \, dx.$$

The convergence of the integrals follows from the fact that $f(x)$ is integrable and $\varphi(t)$ is rapidly decreasing (hence also integrable). By interchanging the order of integration, we obtain $\langle T_{\hat{f}}, \varphi \rangle = \langle \hat{T}_f, \varphi \rangle$, as desired.

Theorem 5.3 (Continuity of the distribution operations). (Compare Theorem 2.3.) Let $\{S_n\}$ and $\{T_n\}$ be sequences of tempered distributions converging to S and T, respectively. Then $S_n + T_n \to S + T$, $cT_n \to cT$ for constant c, $T'_n \to T'$, $\hat{T}_n \to \hat{T}$, and similarly for the other operations defined above.

Proof. Identical to that of Theorem 2.3.

Theorem 5.4 (Identities of calculus). (Compare Theorem 2.4.) Let S and T be tempered distributions, $g(x)$ a C^∞ function which is slowly increasing together with all of its derivatives, and $a \neq 0$, c, respectively, real and complex constants. Then,

$(S + T)' = S' + T'$;

$(cT)' = c \cdot T'$;

$[T(ax)]' = a \cdot T'(ax)$;

$[T(x - a)]' = T'(x - a)$;

$[g(x)T(x)]' = g'(x)T(x) + g(x)T'(x)$.

Proof. We observe that no Fourier transform identities appear here. Then, since every tempered distribution corresponds to a unique general distribution, this result is a corollary of Theorem 2.4.

We turn now to the Fourier transform identities. The first one is so important that we state it separately.

Theorem 5.5 (Fourier Inversion Theorem). Let T be a tempered distribution. Then $\hat{T}$ is a tempered distribution, and

$$T^{\hat{}\check{}} = T^{\check{}\hat{}} = T.$$

Proof. We have already seen, in Theorem 5.1 above, that $\hat{T}$ is a tempered distribution. We use the Fourier Inversion Theorem for open support test functions, proved in Chapter 4: if φ is an open support test function, then so is $\hat{\varphi}$, and

$$\varphi^{\hat{}\check{}} = \varphi^{\check{}\hat{}} = \varphi.$$

Now by definition

$$\langle T^{\hat{}\check{}}, \varphi \rangle = \langle T^{\hat{}}, \varphi^{\check{}} \rangle = \langle T, \varphi^{\check{}\hat{}} \rangle = \langle T, \varphi \rangle.$$

Similarly for $T^{\check{}\hat{}}$.

Remark. Thus, the Fourier Inversion Theorem, originally proved only for the extremely restricted class $\mathcal{S}$, has been extended almost without effort to the dual class $\mathcal{S}'$ of all tempered distributions. This might be a good place to recall that there are many alternative theories of 'generalized functions'. The test function/duality approach of Laurent Schwartz is only one of the many possibilities that have been explored. The preceding proof should give some indication as to why Schwartz's theory is so popular.

Theorem 5.6 (Fourier transform identities). Let S and T be tempered distributions, $a \neq 0$ a real constant, and c, d complex constants. Then,

$$[cS + dT]^{\hat{}} = c \cdot \hat{S} + d \cdot \hat{T};$$
$$[T']^{\hat{}} = 2\pi i \cdot t \cdot \hat{T}(t);$$
$$[xT(x)]^{\hat{}} = (-1/2\pi i)(\hat{T})';$$
$$[T(x-a)]^{\hat{}} = e^{-2\pi i a t}\hat{T}(t);$$
$$[e^{2\pi i a x}T(x)]^{\hat{}} = \hat{T}(t-a);$$
$$[T(ax)]^{\hat{}} = (1/|a|)\hat{T}(t/a);$$
$$[T^*(x)]^{\hat{}} = \bar{\hat{T}}(t), \text{ where } T^*(x) = \overline{T(-x)}.$$

(Of course, the conjugate $\bar{T}$ of a tempered distribution T is defined by $\langle \bar{T}, \varphi \rangle = \overline{\langle T, \bar{\varphi} \rangle}$.)

We observe that the above list contains no identities for convolution or multiplication. These will be treated in Chapter 7.

Proof. Of course, the proofs go by carrying the operations over from the distribution T to the test functions φ. As in Theorem 2.4 in Chapter 2, these calculations can provide a bit of easy relaxation, since the duality causes the operations to jump around in an amusing way, and yet everything comes out right in the end. We do the proof for $[T']\hat{}$, and leave the other cases to the reader. By definition

$$\langle [T']\hat{}, \varphi \rangle = \langle T', \phi \rangle = -\langle T, (\phi)' \rangle,$$

and by the corresponding identity for test functions (Chapter 4, Identity (c)):

$$(\phi)'(t) = -2\pi i[t \cdot \varphi(t)]\hat{}.$$

(Notice how the sign shift in the Fourier transform identities, $(\varphi')\hat{} = 2\pi i t \hat{\phi}(t)$ and $(\hat{\phi})' = -2\pi i[t\varphi(t)]\hat{}$, balances the sign shift in the distribution derivative: $\langle T', \psi \rangle = -\langle T, \psi' \rangle$.) We have

$$-\langle T, (\phi)' \rangle = -\langle T, -2\pi i[t \cdot \varphi(t)]\hat{} \rangle = \langle 2\pi i \hat{T}, t \cdot \varphi(t) \rangle = \langle 2\pi i \cdot t \cdot \hat{T}, \varphi \rangle,$$

as desired.

2 Examples

The examples in this section are of two types. In the first example, we consider a function $f(x)$ which is actually integrable, so that its Fourier transform could be computed by classical methods (contour integration). However, we do not use those methods. As an illustration of distribution technique, we use the generic properties of the Fourier transform. The remaining examples are more striking. Here we consider functions or distributions which are not integrable, so that the Fourier transform cannot be defined classically (as an integral), although it does have a meaning in the distribution sense. In the preceding section we defined the generalized Fourier transform. Here we shall show, in several examples, how to compute it.

As a preliminary to this work, we set down certain basic results which will be used repeatedly. First we observe that the Fourier transform of the delta function,

$$\hat{\delta} = 1,$$

since, for any test function φ,

$$\langle \hat{\delta}, \varphi \rangle = \langle \delta, \hat{\phi} \rangle = \int_{-\infty}^{\infty} e^{-2\pi i t x} \varphi(x)\, dx \quad (\text{at } t = 0)$$

$$= \int_{-\infty}^{\infty} \varphi(x)\, dx = \langle T_1, \varphi \rangle = \langle 1, \varphi \rangle.$$

Remark. From now on, we shall frequently identify a function f with the distribution T_f. This was done above, where we wrote 1 in place of T_1. Earlier on, we were being very careful about logical distinctions. However, the whole point of distribution theory is to create a theory of *generalized functions*. There comes a time when we want to blur the distinction between T_f and f. This is not a matter of indifference; it is quite premeditated.

Theorem 5.7. Let T be a tempered distribution. If $x \cdot T(x) = 0$, then T is a constant multiple of the delta function $\delta(x)$. If $T'(x) = 0$, then T is equivalent to a constant function.

Proof. We begin with T such that $xT(x) = 0$. The proof hinges on the fact that, if $\varphi(x)$ is an open support test function such that $\varphi(0) = 0$, then $\varphi(x)/x$ is again an open support test function: as we recall, this is the '$\varphi(x)/x$ Lemma' which was stated in Chapter 3 and repeated in Chapter 4. Now we deduce: if $\varphi(0) = 0$, then $\langle T, \varphi \rangle = 0$. For $\langle T, \varphi \rangle = \langle xT(x), \varphi(x)/x \rangle = 0$, since $xT(x) = 0$ and $\varphi(x)/x$ is a test function.

The rest is routine. We take a *fixed* test function $\psi \in \mathscr{S}$ such that $\psi(0) = 1$. Then we define Const $= \langle T, \psi \rangle$. Now, for *any* test function φ,

$$[\varphi(x) - \varphi(0)\psi(x)](0) = 0,$$

so that

$$\langle T, \varphi(x) - \varphi(0)\psi(x) \rangle = 0,$$
$$\langle T, \varphi \rangle = \varphi(0)\langle T, \psi \rangle = \text{Const} \cdot \varphi(0) = \text{Const}\langle \delta, \varphi \rangle.$$

Hence $T = \text{Const} \cdot \delta$, as required.

For the case where $T'(x) = 0$, we could give a direct argument, but we prefer to use the Fourier transform. Since $[T']\hat{} = 2\pi i t \cdot \hat{T}(t)$ and $T' = 0$, we deduce that $t \cdot \hat{T}(t) = 0$. Hence, by what we have already shown, $\hat{T}(t) = \text{Const} \cdot \delta(t)$. By the Fourier Inversion Theorem, $T = T\hat{}\hat{} = \text{Const} \cdot 1$ (since $\hat{\delta}(x) = \hat{\delta}(-x) = 1$).   Q.E.D.

Sometimes we only know that T' vanishes on an interval (a, b).

Definition. Let T be a distribution, and let (a, b) be a bounded open interval in $\mathbb{R}^1$. We say that T *vanishes* on (a, b) if $\langle T, \varphi \rangle = 0$ for all test functions $\varphi \in \mathscr{S}$ with support in (a, b). Likewise, for two distributions S and T, we say that $S = T$ on (a, b) if $S - T$ vanishes on (a, b).

The next theorem belongs more properly to the theory of general, rather than tempered, distributions. However, as long as the interval (a, b) is bounded, it makes no difference: every general distribution coincides on (a, b) with a tempered distribution.

Theorem 5.8 (Uniqueness for first order linear ordinary differential equations). Let T be a distribution, $p(x)$ a C^∞ function, and $q(x)$ a continuous function. Suppose that on a bounded open interval (a, b)

$$\frac{dT}{dx} + p(x) \cdot T = q(x).$$

Then, on (a, b),

$$T(x) = e^{-\int_0^x p(u)\,du}\left(\int_0^x e^{\int_0^u p(v)\,dv}\, q(u)\,du + \text{Const}\right).$$

Proof. Of course, our proof must take into account the fact that T is a distribution, and hence the derivative dT/dx might not exist in the classical sense.

Let T_f be the particular (classical) solution to the differential equation given by the above solution formula (with Const $= 0$). Then $T_1 = T - T_f$ satisfies the homogeneous equation $T_1' + p(x)\cdot T_1 = 0$. Let

$$T_2(x) = e^{\int_0^x p(u)\,du}\cdot T_1(x).$$

Then $T_2' = 0$ by the standard differentiation formulas, proved for distributions in Theorem 5.4. We want to show that $T_2 = \text{Const}$ on (a, b).

Now for convenience we replace T_2 by T. Thus the problem reduces to showing that if $T' = 0$ on (a, b), then $T = \text{Const}$ on (a, b). Here, since $(a, b) \neq \mathbb{R}^1$, we cannot use Theorem 5.7. Instead we argue as follows.

Consider any test function φ, supported on (a, b), such that $\int_a^b \varphi(u)\,du = 0$. Then the antiderivative $\theta(x) = \int_{-\infty}^x \varphi(u)\,du$ is a test function supported on (a, b) with $\theta'(x) = \varphi(x)$. Since $T' = 0$ on (a, b),

$$\langle T, \varphi \rangle = \langle T, \theta' \rangle = -\langle T', \theta \rangle = 0.$$

Now choose a *fixed* test function ψ supported in (a, b) and such that $\int_a^b \psi(u)\,du = 1$. Let $\text{Const} = \langle T, \psi \rangle$. Then for *any* test function φ, let $K = \int_a^b \varphi(u)\,du$, so that

$$\int_a^b [\varphi(u) - K\psi(u)]\,du = 0.$$

Hence, as we have already shown,

$$\langle T, \varphi - K\psi \rangle = 0,$$
$$\langle T, \varphi \rangle = K \cdot \langle T, \psi \rangle = \text{Const} \cdot K$$
$$= \text{Const} \cdot \int_a^b \varphi(u)\,du = \text{Const} \cdot \langle 1, \varphi \rangle,$$

as desired.

Example 1. (The function $f(x) = 1/(1 + x^2)$.) This is an integrable function, and its Fourier transform, as defined classically, is

$$\hat{f}(t) = \int_{-\infty}^\infty \frac{e^{-2\pi itx}}{1 + x^2}\,dx = \int_{-\infty}^\infty \frac{\cos 2\pi tx}{1 + x^2}\,dx,$$

since the imaginary part vanishes by symmetry. If we did the integral (e.g., by contour integration) we would find: $\hat{f}(t) = \pi e^{-2\pi|t|}$. We prefer to use distribution theory; the fact that the result is classically (as well as distribution-theoretically) valid follows from the Consistency Theorem.

Since $f(x) = 1/(1 + x^2)$, the product $(1 + x^2)f(x) = 1$. Since $\hat{\delta} = \check{\delta} = 1$, we have, by the Fourier Inversion Theorem,

$$\hat{1} = \check{1} = \delta.$$

Since $[xf(x)]\hat{} = (-1/2\pi i)[\hat{f}(t)]'$, we obtain

$$[(1 + x^2)f(x)]\hat{} = \hat{f}(t) - (1/4\pi^2)\frac{d^2}{dt^2}\hat{f}(t) = \delta(t).$$

Now $\delta(t)$ vanishes except at $t = 0$, and so for $t \neq 0$ we see that $\hat{f}$ satisfies the differential equation

$$\hat{f}'' - (4\pi^2)\hat{f} = 0.$$

This equation has the general solution $\hat{f}(t) = Ae^{2\pi t} + Be^{-2\pi t}$
(Note: this is justified by Theorem 5.8, applied to any open interval (a, b) not containing the origin. For the second order differential operator $(D^2 - 4\pi^2)$ factors into $(D - 2\pi)(D + 2\pi)$.)

Since $f(x)$ is an integrable function, $\hat{f}(t)$ is bounded; hence

$$\hat{f}(t) = \begin{cases} Ae^{-2\pi t} & \text{if } t > 0 \\ Be^{2\pi t} & \text{if } t < 0, \end{cases}$$

for some constants A and B. Since $f(x)$ is real valued and an even function, $A = B$. Finally, we consider the point $t = 0$. Here

$$\hat{f}''(t) = -4\pi^2\,\delta(t) + \text{(an ordinary function)}.$$

Now the first derivative $\hat{f}'(t)$ jumps from $2\pi A$ to $-2\pi A$ as t moves from the left to the right side of $t = 0$. Hence the second derivative

$$\hat{f}''(t) = -4\pi A \cdot \delta(t) + \text{(an ordinary function)}.$$

Comparing the last two displayed equations, we have $A = \pi$. Hence,

$$f(t) = \pi e^{-2\pi|t|}.$$

Example 2. (The function x^n and related functions.) Of course, x^n is not integrable on $\mathbb{R}^1$. Nevertheless, it gives a tempered distribution, and so it must have a Fourier transform. This we now compute.

We have already seen that $1\hat{} = \delta(t)$, and from the identity $[xT(x)]\hat{} = (-1/2\pi i)\hat{T}'$, we deduce

$$[x^n]\hat{} = (-1/2\pi i)^n \cdot \delta^{(n)}(t).$$

From the Fourier Inversion Theorem, together with the fact that $\hat{T}(x) = \check{T}(-x)$, we deduce

$$[\delta^{(n)}(x)]\hat{} = (2\pi it)^n.$$

(As an alternative derivation of the above, we could have started with $\hat{\delta} = 1$ and used the identity $[T']\hat{} = 2\pi it \cdot \hat{T}(t)$.)

From the various identities in Theorem 5.6, we automatically derive a number of related formulas, e.g.

$$[(x-a)^n]\hat{} = e^{-2\pi iat} \cdot (-1/2\pi i)^n \cdot \delta^{(n)}(t),$$

$$[e^{2\pi iax} \cdot x^n]\hat{} = (-1/2\pi i)^n \delta^{(n)}(t-a),$$

$$[\delta^{(n)}(x-a)]\hat{} = e^{-2\pi iat} \cdot (2\pi it)^n.$$

(Note that the sign change in the exponent in the last two formulas, corresponding to the distinction between $\hat{}$ and $\check{}$.)

Example 3. (The pseudofunction $1/x^n$, the function $|x| \cdot x^{n-1}$, and the Heaviside function $H(x)$.) The Heaviside function is defined by

$$H(x) = \begin{cases} \frac{1}{2} & \text{for } x > 0 \\ -\frac{1}{2} & \text{for } x < 0. \end{cases}$$

We observe that $H(x)$ differs from the step function introduced in Chapter 1 by an additive constant of $\frac{1}{2}$. We prefer this variant of Heaviside's function because it is *odd*, i.e. $H(-x) = -H(x)$. The most important property of $H(x)$ is

$$H'(x) = \delta(x).$$

(A rigorous proof of this intuitively obvious fact was given in Chapter 1.)

Now we compute the Fourier transform of $H(x)$. Since $H'(x) = \delta(x)$, and $\hat{\delta} = 1$, the identity $[T']\hat{} = (2\pi it)\hat{T}(t)$ yields $\hat{H}(t) = 1/2\pi it$. Actually, this was a touch glib. After Theorem 5.7, we should compute as follows: $H' = \delta$, whence $[H']\hat{} = 1$, whence $(2\pi it)\hat{H}(t) = 1$, whence $\hat{H}(t) = (1/2\pi it) + \text{Const} \cdot \delta(t)$ for some (so far unspecified) value of the constant. To determine Const, we observe that $H(x)$ is a real valued odd function; thus $\hat{H}(t)$ is purely imaginary and odd. Since $\delta(t)$ is an even distribution, Const $= 0$. We have proved

$$[H(x)]\hat{} = \frac{1}{2\pi it}.$$

(Of course, $1/t$ is a pseudofunction, as defined in Chapter 3.)

Now we are in a position to deal with the functions $|x| \cdot x^{n-1}$. Before we begin, we observe that the half-sided function

$$h_n(x) = \begin{cases} x^n & \text{for } x \geqslant 0 \\ 0 & \text{for } x < 0 \end{cases}$$

can be expressed as $\frac{1}{2}[x^n + |x|x^{n-1}]$. We already know the Fourier transform of x^n (Example 2); thus, in computing the transform of $|x| \cdot x^{n-1}$, we are computing that of $h_n(x)$ as well.

Now $|x| \cdot x^{n-1} = 2x^n H(x)$. From $[xT(x)]\hat{} = (-1/2\pi i)\hat{T}'$ we deduce that $[|x| \cdot x^{n-1}]\hat{} = 2(-1/2\pi i)^n (d/dt)^n [1/2\pi it]$. Hence, since $(d/dt)^n(1/t) = (-1)^n \cdot n! \cdot (1/t^{n+1})$ (see Chapter 3),

$$[|x| \cdot x^{n-1}]\hat{} = 2 \cdot (1/2\pi i)^{n+1} \cdot n! \cdot \frac{1}{t^{n+1}}.$$

In particular,

$$|x|\hat{\ } = (-1/2\pi^2)\frac{1}{t^2}.$$

Now using the Fourier Inversion Theorem (and, as usual, keeping track of the fact that $\hat{T}(x) = \check{T}(-x)$), we obtain

$$\left[\frac{1}{x^n}\right]\hat{\ } = (-1)^n \cdot \frac{1}{2} \cdot \frac{(2\pi i)^n}{(n-1)!} \cdot |t| \cdot t^{n-2},$$

whence, in particular,

$$\left[\frac{1}{x}\right]\hat{\ } = -\pi i \cdot \text{sign}(t) = -2\pi i \cdot H(t).$$

Example 4. (The pseudofunction $1/|x|$, the function $\log|x|$, and related functions.) First we must define what we mean by $1/|x|$, since this function is not integrable near $x = 0$, nor does it give a Cauchy principle value. We define

$$\log^* x = \begin{cases} \log x & \text{for } x > 0 \\ -\log(-x) & \text{for } x < 0. \end{cases}$$

Thus $\log^* x = \text{sign}(x) \cdot \log|x|$. Now $\log^* x$ is integrable near $x = 0$, and it is slowly increasing as $|x| \to \infty$; hence it corresponds to a tempered distribution. We define

$$\frac{1}{|x|} = \frac{d}{dx}(\log^* x).$$

One readily verifies, as expected, that

$$x \cdot \frac{1}{|x|} = \text{sign}(x) = 2 \cdot H(x).$$

As in Examples 2 and 3, this leads by simple manipulations to a formula for $[1/|x|]\hat{\ }$. Again we use the generic identity $[xT(x)]\hat{\ } = (-1/2\pi i)\hat{T}'$. Let $T = 1/|x|$. Since $x \cdot T(x) = 2H(x)$, $(-1/2\pi i)\hat{T}'(t) = 2\hat{H}(t) = (1/\pi i)(1/t)$ by Example 3. Following Chapter 3 we see that $d/dt(\log|t|) = 1/t$. Hence, by Theorem 5.7, $\hat{T}(t) = -2\log|t| + \text{Const}$. We have proved

$$\left[\frac{1}{|x|}\right]\hat{\ } = -2\log|t| + C,$$

where C is some constant. Actually,

$$C = -2(\gamma + \log 2\pi),$$

where γ denotes Euler's constant.

We shall work out the value of C below. Putting it aside for now, let us deduce the obvious consequences of our result.

First we define

$$\frac{1}{|x| \cdot x^{n-1}} = (-1)^{n-1} \frac{1}{(n-1)!} \left(\frac{d}{dx}\right)^{n-1} \frac{1}{|x|}.$$

As in Example 3, the half-sided function

$$h_{-n}(x) = \begin{cases} 1/x^n & \text{for } x > 0 \\ 0 & \text{for } x < 0 \end{cases}$$

can be expressed as the average of $1/x^n$ and $1/|x|x^{n-1}$. So again, in dealing with $1/|x|x^{n-1}$, we are also taking care of $h_{-n}(x)$. Now, for $1/|x|x^{n-1}$, we use $[T']\hat{} = (2\pi i t)\hat{T}(t)$ to obtain

$$\left[\frac{1}{|x| \cdot x^{n-1}}\right]\hat{} = \frac{(-2\pi i t)^{n-1}}{(n-1)!} [-2\log|t| + C].$$

To obtain the Fourier transform of $\log|x|$, we take the result already obtained for $[1/|x|]\hat{}$ and apply the Fourier Inversion Theorem:

$$[\log|x|]\hat{} = \frac{-1}{2|t|} + \frac{C}{2} \cdot \delta(t),$$

with the same constant C as above.

Similarly, since $(d/dx)(\log^* x) = 1/|x|$, and $\log^* x$ is an odd function,

$$[\log^* x]\hat{} = \frac{-2\log|t| + C}{2\pi i t}.$$

Now we come to the computation of C. This is by far the most difficult part of the whole exercise. Let $T(x) = 1/|x|$, so that $\hat{T}(t) = -2\log|t| + C$. As in Chapter 4, we use that Gaussian test function which is its own Fourier transform:

$$\varphi(x) = e^{-\pi x^2}, \qquad \phi(t) = e^{-\pi t^2}.$$

Since $\langle \hat{T}, \varphi \rangle = \langle T, \phi \rangle$ and $1/|x| = (d/dx)(\log^* x)$,

$$\left\langle -2\log|t| + C, \varphi(t) \right\rangle = \left\langle \frac{1}{|x|}, \phi(x) \right\rangle = -\langle \log^* x, \phi(x)' \rangle.$$

The left side is equal by definition to

$$\int_{-\infty}^{\infty} (-2\log|t| + C)e^{-\pi t^2} dt = 2 \int_0^{\infty} (-2\log|t| + C)e^{-\pi t^2} dt,$$

while similarly the right side is

$$-2 \int_0^{\infty} \log x(-2\pi x)e^{-\pi x^2} dx.$$

Now we utilize some facts about the gamma function. (See, e.g., [WHW].) In the first place, all of the above integrals are of the type covered by the well known formulas:

$$\int_0^\infty x^{2m-1} e^{-ax^2}\, dx = \frac{1}{2} \frac{\Gamma(m)}{a^m},$$

$$\int_0^\infty \log x \cdot x^{2m-1} e^{-ax^2}\, dx = \frac{1}{4}\left[\frac{\Gamma'(m)}{a^m} - \frac{\Gamma(m)\log a}{a^m}\right].$$

(The second formula follows from the first upon taking d/dm.) Thus we can compute the various integrals above: it will be obvious from the results which values of m and a have been used.

$$-4\int_0^\infty \log t \cdot e^{-\pi t^2}\, dt = -\frac{\Gamma'(1/2)}{\pi^{1/2}} + \frac{\Gamma(1/2)\log \pi}{\pi^{1/2}}.$$

$$2C\int_0^\infty e^{-\pi t^2}\, dt = C\cdot\Gamma(1/2)/\pi^{1/2}.$$

$$4\pi\int_0^\infty \log x \cdot x e^{-\pi x^2}\, dx = \pi\left[\frac{\Gamma'(1)}{\pi} - \frac{\Gamma(1)\log \pi}{\pi}\right].$$

Now all of the values of $\Gamma(x)$ and $\Gamma'(x)$ which appear in these formulas are known explicitly:

$$\Gamma(1) = 1, \qquad \Gamma'(1) = -\gamma, \qquad \Gamma\left(\frac{1}{2}\right) = \pi^{1/2}, \qquad \frac{\Gamma'}{\Gamma}\left(\frac{1}{2}\right) = -\gamma - 2\log 2.$$

(See [WHW], chap. 12; cf. especially Problem 3. This problem follows easily from the Legendre Duplication Formula given earlier in the chapter.)

Since the sum of the first two integrals above equals the third, we have

$$-(-\gamma - 2\log 2) + \log \pi + C = -\gamma - \log \pi,$$
$$C = -2\gamma - 2\log 2\pi.$$

3 Bochner's Theorem

Bochner's Theorem gives the relationship, via the Fourier transform, between positive and positive definite distributions. These we now define.

Definition. Let T be a tempered distribution. We say that T is *positive* if $\langle T, \varphi\rangle \geqslant 0$ for every test function $\varphi \in \mathscr{S}$ with $\varphi \geqslant 0$.

Before stating the next definition, we recall the formulas for convolution $(\varphi * \psi)$ and involution (φ^*), as they apply to test functions:

$$(\varphi * \psi)(x) = \int_{-\infty}^\infty \varphi(x - t)\psi(t)\, dt,$$
$$\varphi^*(x) = \overline{\varphi(-x)}.$$

We recall that (Chapter 4):
$$(\varphi * \psi)\hat{} = \hat{\varphi}\hat{\psi},$$
$$(\varphi^*)\hat{} = \overline{\hat{\varphi}}.$$
From these formulas we deduce
$$(\varphi * \varphi^*)(x) = \int_{-\infty}^{\infty} \varphi(x+t)\overline{\varphi(t)}\, dt,$$
$$(\varphi * \varphi^*)\hat{} = |\hat{\varphi}|^2.$$
The expression $(\varphi * \varphi^*)(x)$ is sometimes called the *autocorrelation function* of φ.

Definition. Let T be a tempered distribution. We say that T is *positive definite* if $\langle T, \varphi * \varphi^* \rangle \geqslant 0$ for every test function $\varphi \in \mathcal{S}$.

Theorem 5.9 (Bochner's Theorem). Let T be a tempered distribution. Then T is positive definite if and only if its Fourier transform $\hat{T}$ is positive.

Remarks. We could just as well give the condition that the *inverse* Fourier transform $\check{T}$ is positive; for $\check{T}(x) = \hat{T}(-x)$ is positive if and only if $\hat{T}(x)$ is. Likewise, by the Fourier Inversion Theorem, Bochner's Theorem could be stated: a tempered distribution T is positive if and only if $\hat{T}$ is positive definite.

Proof. We begin with the 'if' part. Suppose $\hat{T}$ is positive. Then
$$\langle T, \varphi * \varphi^* \rangle = \langle T^{\hat{}\check{}}, \varphi * \varphi^* \rangle \quad \text{(Fourier Inversion Theorem)}$$
$$= \langle \hat{T}, (\varphi * \varphi^*)\check{} \rangle \quad \text{(definition of } S\check{})$$
$$= \langle \hat{T}, |\check{\varphi}|^2 \rangle \quad \text{(by the above formulas)}.$$
Since $\hat{T}$ is positive, and $|\check{\varphi}|^2 > 0$, we have $\langle T, \varphi * \varphi^* \rangle = \langle \hat{T}, |\check{\varphi}|^2 \rangle \geqslant 0$, so that T is positive definite, as desired.

The 'only if' part is a little harder. The difficulty is that we have to take a square-root, and the function $x^{1/2}$ is not C^{∞} at $x = 0$. We use an approximation argument.

Lemma. Let $\varphi \in \mathcal{S}$, $\varphi \geqslant 0$. Then there exists a sequence of functions $\theta_n \in \mathcal{D}$, $\theta_n \geqslant 0$, such that $\theta_n^2 \to \varphi$ in $\mathcal{S}$.

Note. $\theta_n \in \mathcal{D}$ has the following useful consequences. Suppose that T is a tempered distribution such that $\langle T, \varphi \rangle \geqslant 0$ for all $\varphi \geqslant 0$, $\varphi \in \mathcal{D}$ ($\mathcal{D}$ not $\mathcal{S}$). Then $\langle T, \varphi \rangle \geqslant 0$ for all $\varphi \geqslant 0$, $\varphi \in \mathcal{S}$; i.e. T is positive on $\mathcal{S}$.

Proof of lemma. We use the Mesa Function Lemma introduced at the beginning of section 1 (p. 61). Recall that a mesa function ψ is a function which is C^{∞} with compact support, $0 \leqslant \psi \leqslant 1$, and $\psi(x) \equiv 1$ on some neighborhood of $x = 0$. We let $\psi_a(x) = \psi(x/a)$. Then the lemma tells us that, for any $\varphi \in \mathcal{S}$, $\psi_a \varphi \to \varphi$ in $\mathcal{S}$ as $a \to \infty$.

We observe that, if ψ is a mesa function, then so is ψ^2. Hence we also have $\psi_a^2\varphi \to \varphi$ in $\mathscr{S}$ as $a \to \infty$.

Take any constant $\varepsilon > 0$. Since $\varphi(x) \geqslant 0$, the function $\varphi(x) + \varepsilon$ is bounded below by ε. Hence we can now take the square-root: $[\varphi(x) + \varepsilon]^{1/2}$ is a C^∞ function. Let

$$\theta_{a,\varepsilon}(x) = \psi_a(x)[\varphi(x) + \varepsilon]^{1/2}.$$

Then $\theta_{a,\varepsilon}$ is C^∞ with compact support, i.e. $\theta_{a,\varepsilon} \in \mathscr{D}$. Now

$$(\theta_{a,\varepsilon})^2 = \psi_a^2(\varphi + \varepsilon) = \psi_a^2\varphi + \varepsilon\psi_a^2.$$

By the mesa function property, $\psi_a^2\varphi \to \varphi$ in $\mathscr{S}$ as $a \to \infty$. For any given a, we can choose ε (depending on a) so that $\varepsilon\psi_a^2$ approximates zero in $\mathscr{S}$. Hence,

$$\lim_{a \to \infty} (\theta_{a,\varepsilon(a)})^2 = \varphi.$$

Finally, we can create a sequence θ_n by choosing any sequence of a-values approaching infinity, e.g. by setting $a_n = n$, $\varepsilon_n = \varepsilon(a_n)$.

Proof of Bochner's Theorem, completed. We must show the 'only if' part: that when T is positive definite, $\hat{T}$ is positive.

By the lemma, to show $\hat{T}$ is positive, it suffices to show that $\langle \hat{T}, \theta^2 \rangle \geqslant 0$ for every $\theta \geqslant 0$ in $\mathscr{D}$. Actually, we shall show more: that $\langle \hat{T}, |\varphi|^2 \rangle \geqslant 0$ for all $\varphi \in \mathscr{S}$.

Now the 'only if' part follows via the same chain of calculations used above for the 'if' part. There we found

$$\langle T, \varphi * \varphi^* \rangle = \langle \hat{T}, |\check{\phi}|^2 \rangle,$$

whence

$$\langle \hat{T}, |\varphi|^2 \rangle = \langle T, \hat{\phi} * \hat{\phi}^* \rangle,$$

upon substituting $\hat{\phi}$ for φ, and using $\varphi^{\wedge\vee} = \varphi$.

If T is positive definite, then $\langle T, \hat{\phi} * \hat{\phi}^* \rangle \geqslant 0$ for all $\varphi \in \mathscr{S}$, whence $\langle \hat{T}, |\varphi|^2 \rangle \geqslant 0$, as desired.

We conclude this section with an example illustrating Bochner's Theorem. Naturally we prefer an example which requires distribution theory – i.e. where the objects considered are not describable as ordinary functions.

Example. We begin with the pseudofunction $1/x^2$, as defined in Chapter 3. Recall the definition:

$$(1/x^2) = -(d/dx)(1/x),$$

where $1/x$ has the 'natural' definition

$$\langle 1/x, \varphi(x) \rangle = \int_{-\infty}^{\infty} (1/x)\varphi(x)\,dx \quad \text{(Cauchy principal value)}.$$

Now $1/x^2$ is not a function (in the classical sense) because of its wild behavior near $x = 0$. That is, the action of $1/x^2$ on a test function $\varphi(x)$ is *not* given by $\langle 1/x^2, \varphi(x) \rangle = \int_{-\infty}^{\infty} (1/x^2)\varphi(x)\, dx$. In fact, as we saw in Chapter 3, $1/x^2$ is *not positive*: the value $\langle 1/x^2, \varphi \rangle$ can be negative even though $\varphi \geqslant 0$. Of course, $1/x^2$ is not everywhere negative either – obviously it behaves like a positive distribution over any interval (a, b) which excludes the point $x = 0$.

(To give a crude intuitive picture: the pseudofunction $1/x^2$ is positive for all $x \neq 0$, but includes a negatively infinite mass situated at the point $x = 0$. This is perhaps not surprising if one thinks of the definition $(1/x^2) = (d/dx)(-1/x)$, and of the fact that the graph of $-1/x$ takes a 'downward jump of $-\infty$' at $x = 0$.)

However, $1/x^2$ does have one attractive property which is easily described. Its negative, $-1/x^2$, is positive definite. For, as we saw in Example 3 of the preceding section, the Fourier transform of $(-1/x^2)$ is $2\pi^2 \cdot |t|$. Now, $2\pi^2 \cdot |t|$ is a perfectly ordinary slowly increasing function. (It is not integrable, but in distribution theory that does not matter.) Clearly $2\pi^2 \cdot |t|$ is positive – i.e. is a positive distribution – since, for all $\varphi \geqslant 0$,

$$\langle 2\pi^2 \cdot |t|, \varphi(t) \rangle = \int_{-\infty}^{\infty} 2\pi^2 |t| \varphi(t)\, dt \geqslant 0.$$

Hence, by Bochner's Theorem, $-1/x^2$ is positive definite.

Remarks (concerning measures). For readers who know measure theory, we mention the following facts. Every positive distribution is given by a measure. More precisely:

Theorem. Let T be a positive tempered distribution. Then there exists a positive measure μ such that

$$\langle T, \varphi \rangle = \int_{-\infty}^{\infty} \varphi(x)\, d\mu(x)$$

for all test functions $\varphi \in \mathscr{S}$.

We do not need this result and shall not prove it.

Finally, we observe that an arbitrary distribution (not necessarily positive) need *not* be given by a signed measure. For example, the dipole distribution $-\delta'(x)$ is not given by any signed measure.

4 Fourier series

We begin by recalling the key facts in the classical theory of Fourier series. These series are used to represent periodic functions, i.e. functions $f(x)$ such that, for some fixed number $T > 0$, $f(x + T) = f(x)$ identically for all x. The number T is called the *period*. Since, if $f(x)$ has period T then $f(xT)$ has period 1, there is no loss of generality in assuming that the period is 1, and we shall do so in what follows.

(In applications, it is frequently useful to have the formulas for a general period T. For the convenience of the reader, we shall list the key formulas for period T in the next 'Examples' section.)

For functions $f(x)$ of period 1, the Fourier series is conventionally given in the form

$$f(x) = \tfrac{1}{2}a_0 + \sum_{n=1}^{\infty} (a_n \cos 2\pi nx + b_n \sin 2\pi nx),$$

where

$$a_n = 2 \int_0^1 f(x) \cos 2\pi nx \, dx, \qquad b_n = 2 \int_0^1 f(x) \sin 2\pi nx \, dx.$$

(The factor '2' instead of the usual '$1/\pi$' occurs because we have chosen to use period 1 instead of period 2π.)

For theoretical purposes, the complex form of Fourier series is easier to deal with. This is

$$f(x) = \sum_{n=-\infty}^{\infty} c_n e^{2\pi inx},$$

where

$$c_n = \int_0^1 f(x) e^{-2\pi inx} \, dx.$$

For convenience, we list the transition formulas:

$$a_n = c_n + c_{-n}, \qquad b_n = i(c_n - c_{-n});$$
$$c_n = (a_n - ib_n)/2, \qquad c_{-n} = (a_n + ib_n)/2.$$

In the complex form, we observe a strong analogy with Fourier transforms: the formula $c_n = \int_0^1 f(x) e^{-2\pi inx} \, dx$ corresponding to the Fourier transform, and the series $f(x) = \sum_{n=-\infty}^{\infty} c_n e^{2\pi inx}$ corresponding to the inverse Fourier transform. Let us push this analogy a little further. The extensions may not seem obvious at first glance, but, as this section will show, they embody perfect 20/20 hindsight.

The Fourier coefficients c_n are defined only for integer values of n. This might suggest that the Fourier transform $\hat{f}(t)$ is a series of delta functions: these delta functions should be located at the integers $0, \pm 1, \pm 2, \ldots$, and the delta function at $t = n$ should be multiplied by the coefficient c_n. That is, we conjecture that

$$\hat{f}(t) = \sum_{n=-\infty}^{\infty} c_n \cdot \delta(t - n),$$

where c_n is the nth Fourier coefficient of f. Then the inverse Fourier transform is given formally by the integral

$$f(x) = \int_{-\infty}^{\infty} e^{2\pi i x t} \hat{f}(t) \, dt$$

$$= \sum_{n=-\infty}^{\infty} \int_{-\infty}^{\infty} e^{2\pi i x t} \cdot c_n \cdot \delta(t-n) \, dt$$

$$= \sum_{n=-\infty}^{\infty} c_n e^{2\pi i n x},$$

since the delta function $\delta(t-n)$ picks out the value $t = n$ in the integrand $e^{2\pi i x t}$.

Of course, this fails to be rigorous in several respects: we have interchanged an integration and summation without careful proof; and our application of $\delta(t-n)$ to $e^{2\pi i x t}$ would only be valid if $e^{2\pi i x t}$ were a test function, which it is not. Nevertheless, one suspects that this can all be straightened out. More important is the fact that our original conjecture – that $\hat{f}(t)$ is a series of delta functions $c_n \cdot \delta(t-n)$ weighted by the Fourier coefficients c_n – has led in a natural way to the standard Fourier series expansion of f.

In the following subsections, we shall prove all of this rigorously. Moreover, we shall prove it in the maximum generality that one could expect – for arbitrary tempered distributions of period 1.

The key theorem in the distribution-theoretic treatment of Fourier series is the following.

Theorem 5.10. Let T be a tempered distribution which is periodic with period 1, i.e. such that $T(x+1) = T(x)$. Then the Fourier transform $\hat{T}$ has the representation

$$\hat{T}(t) = \sum_{n=-\infty}^{\infty} c_n \cdot \delta(t-n),$$

where the c_n are complex constants, and the series converges to $\hat{T}(t)$ in the sense of tempered distributions (i.e. in $\mathscr{S}'$).

As an immediate corollary we have:

Theorem 5.11. With the assumptions and notation of Theorem 5.10,

$$T(x) = \sum_{n=-\infty}^{\infty} c_n \cdot e^{2\pi i n x},$$

where the convergence is in the sense of tempered distributions.

The proof of Theorem 5.10 is long and hard; it will occupy most of the remainder of this section. Here we dispose of:

Proof of Theorem 5.11 (assuming Theorem 5.10). We simply apply the Fourier Inversion Theorem to the identity given in Theorem 5.10. Since the inverse Fourier transform is continuous on $\mathscr{S}'$, we are justified in operating on the series term by term. Thus:

$$T(x) = T^{\hat{}\check{}}(x) = \sum_{n=-\infty}^{\infty} c_n[\delta(t-n)]^{\check{}}$$

$$= \sum_{n=-\infty}^{\infty} c_n \cdot e^{2\pi i n x},$$

since $[\delta(t-n)]^{\check{}} = e^{2\pi i n x}$.

Results. Of course, we call c_n the 'nth Fourier coefficient' of $T(x)$. This is a definition, not a theorem, since we have no previous definition of 'Fourier coefficient' for a general periodic distribution T. On the other hand, we should expect:

Theorem 5.12 (Consistency Theorem). Let $f(x)$ be a periodic function, $f(x) = f(x+1)$, such that $f(x)$ is integrable on the interval $\{0 \leqslant x \leqslant 1\}$. Let c_n denote the nth Fourier coefficient of the distribution T_f, as determined by Theorem 5.10. Then:

$$c_n = \int_0^1 e^{-2\pi i n x} f(x) \, \mathrm{d}x.$$

These are the main results. Two others of a more technical nature (Theorems 5.13 and 5.14) will be given as we proceed.

Outline of the proof of Theorem 5.10. We begin with an arbitrary tempered distribution T such that $T(x+1) = T(x)$. From Theorem 5.6 we have the identity $[T(x+1)]^{\hat{}} = e^{2\pi i t} \cdot \hat{T}(t)$, whence

$$(e^{2\pi i t} - 1) \cdot \hat{T}(t) = 0.$$

Now we recall the result (Theorem 5.7) that if $x \cdot T(x) = 0$, then $T(x)$ is a constant multiple of the delta function. Here the multiplier x vanishes only at $x = 0$, which is where the delta function is located. (For, on a closed interval where $x \neq 0$, we should expect that we could divide by x, and from the equation $x \cdot T(x) = 0$ deduce that $T(x) = 0$ on this interval.) Now examine the equation involving $\hat{T}(t)$ above. There, instead of the multiplier t, we have the multiplier $(e^{2\pi i t} - 1)$. Whereas t vanished only at $t = 0$, $(e^{2\pi i t} - 1)$ vanishes at $t = 0, \pm 1, \pm 2, \ldots$. This strongly suggests, by analogy with Theorem 5.7, that $\hat{T}(t)$ should consist of constant multiples of delta functions located at the points $t = 0, \pm 1, \pm 2, \ldots$. That is, $\hat{T}(t)$ should have the form $\hat{T}(t) = \sum c_n \cdot \delta(t-n)$ for some sequence of constants $\{c_n\}$. Hopefully, the series $\sum c_n \cdot \delta(t-n)$ will converge in $\mathscr{S}'$. This, essentially, is Theorem 5.10.

So far we have only an analogy with Theorem 5.7. In the next subsection, we shall turn this analogy into a fact accomplished. For convenience, we shall restrict our attention to test functions $\varphi \in \mathcal{D}$ instead of $\varphi \in \mathcal{S}$.

Proof of Theorem 5.10 for $\varphi \in \mathcal{D}$. As stated above, we begin by restricting our attention to $\varphi \in \mathcal{D}$; the extension to $\varphi \in \mathcal{S}$ will come later.

Lemma. Let $\varphi \in \mathcal{D}$ be a test function which vanishes at all integer points, i.e. $\varphi(t) = 0$ for $t = 0, \pm 1, \pm 2, \ldots$. Then the function

$$\psi(t) = \frac{\varphi(t)}{e^{2\pi i t} - 1} \in \mathcal{D}.$$

Proof. Clearly ψ has compact support. The problem is to show that ψ is C^{∞}. The only difficulty occurs at the points $t = 0, \pm 1, \pm 2, \ldots$, where $(e^{2\pi i t} - 1)$ is zero. Since $(e^{2\pi i t} - 1)$ is periodic with period 1, the problem is essentially identical for each integer k, and there is no loss of generality in considering only the point $t = 0$.

Now, previously we have seen that, if $\varphi(t) = 0$, then $\varphi(t)/t \in C^{\infty}$ (the '$\varphi(x)/x$ Lemma' given in Chapter 3 and again in Chapter 4).

On the other hand, $(e^{2\pi i t} - 1) = 2i \cdot e^{\pi i t}[(e^{\pi i t} - e^{-\pi i t})/2i] = 2i \cdot e^{\pi i t} \cdot \sin \pi t$. Hence,

$$\frac{\varphi(t)}{e^{2\pi i t} - 1} = \frac{\varphi(t)}{t} \cdot \frac{t}{\sin \pi t} \cdot (2i)^{-1} e^{-\pi i t}.$$

Now, as is well known, the function $(\sin \pi t)/t$ is C^{∞} and takes the value π at $t = 0$; hence, $t/(\sin \pi t)$ is also C^{∞} near $t = 0$. We have seen that $\varphi(t)/t$ is C^{∞}, and of course $(2i)^{-1} e^{-\pi i t}$ is C^{∞}. Hence, $\varphi(t)/[e^{2\pi i t} - 1]$ is C^{∞} near $t = 0$, as desired.

Lemma. Let T be a tempered distribution with period 1, i.e. $T(x + 1) = T(x)$. Let $\varphi \in \mathcal{D}$ be a test function which vanishes at all integer points. Then

$$\langle \hat{T}, \varphi \rangle = 0.$$

Proof. As noted above: since $T(x + 1) = T(x)$, and since $[T(x + 1)]\hat{} = e^{2\pi i t} \cdot \hat{T}(t)$, we have

$$(e^{2\pi i t} - 1) \cdot \hat{T}(t) = 0.$$

By the previous lemma, since $\varphi(t)$ vanishes at the integers, the function $\psi(t) = \varphi(t)/(e^{2\pi i t} - 1)$ is a test function. Then $\varphi(t) = (e^{2\pi i t} - 1)\psi(t)$, and we deduce

$$\langle \hat{T}, \varphi \rangle = \langle \hat{T}, (e^{2\pi i t} - 1)\psi(t) \rangle = \langle (e^{2\pi i t} - 1)\hat{T}(t), \psi \rangle = 0,$$

since the distribution $(e^{2\pi i t} - 1)\hat{T}(t) = 0$.   Q.E.D.

The next lemma already gives us a substantial part of Theorem 5.10.

Lemma. Let T be a tempered distribution with $T(x + 1) = T(x)$. Then there is a sequence of constants c_n $(n = 0, \pm 1, \pm 2, \ldots)$ depending on T such that, for all $\varphi \in \mathscr{D}$,

$$\langle \hat{T}, \varphi \rangle = \sum_{n=-\infty}^{\infty} \langle c_n \cdot \delta(t - n), \varphi(t) \rangle = \sum_{n=-\infty}^{\infty} c_n \cdot \varphi(n).$$

Notes. Since φ has compact support, the above is a finite sum, and there is no question about its convergence. Once the above is proved, the extension from $\varphi \in \mathscr{D}$ to $\varphi \in \mathscr{S}$ is the only thing lacking for a proof of Theorem 5.10.

Proof. Fix a function $\theta \in \mathscr{D}$ with support on $[-\frac{1}{2}, \frac{1}{2}]$ and such that $\theta(0) = 1$. For example, we can take

$$\theta(x) = \begin{cases} e^{-[x^2/(1-4x^2)]} & \text{for } |x| < \frac{1}{2} \\ 0 & \text{for } |x| \geqslant \frac{1}{2}. \end{cases}$$

We hold θ fixed, and consider arbitrary functions $\varphi \in \mathscr{D}$. First we define

$$c_n = \langle \hat{T}, \theta(t - n) \rangle, \qquad n = 0, \pm 1, \pm 2, \ldots.$$

Now we shall prove that, with these c_n, the desired formula holds for all $\varphi \in \mathscr{D}$. For any $\varphi \in \mathscr{D}$, define $\psi = \psi_\varphi$ by

$$\psi(t) = \varphi(t) - \sum_{n=-\infty}^{\infty} \varphi(n) \cdot \theta(t - n).$$

(Of course, since φ has compact support, the sum is finite.)

Since $\theta(0) = 1$ and support $(\theta) \subseteq [-\frac{1}{2}, \frac{1}{2}]$, the function ψ vanishes at every integer. Hence, by the previous lemma,

$$\langle \hat{T}, \psi \rangle = 0.$$

But, by definition of c_n,

$$\langle \hat{T}, \psi \rangle = \langle \hat{T}, \varphi \rangle - \sum_{n=-\infty}^{\infty} \varphi(n) \cdot c_n.$$

Combining the last two displayed equations gives the desired result.

The extension to $\varphi \in \mathscr{S}$. The key step is to show that the Fourier coefficients c_n do not grow too rapidly. In fact, these coefficients are 'slowly increasing', i.e. they are dominated by a fixed power of n. This power, however, may be different for different periodic distributions T.

Theorem 5.13. Let T be a periodic tempered distribution, with $T(x + 1) = T(x)$, and let $\{c_n\}$ be its sequence of Fourier coefficients; as defined above. Then there exists a constant > 0 and an integer N (depending on T) such that

$$|c_n| < \text{Const} \cdot (1 + |n|)^N \text{ for all } n.$$

Conversely, every sequence $\{c_n\}$ with $|c_n| \leqslant \text{Const} \cdot (1 + |n|)^N$ for some Const and N, corresponds in this way to some periodic distribution T.

Proof. The converse part is trivial. For if $|c_n| \leqslant \text{Const} \cdot (1 + |n|)^N$, then the series

$$\hat{T}(t) = \sum_{n=-\infty}^{\infty} c_n \cdot \delta(t - n)$$

converges in $\mathscr{S}'$. Namely, by the definition of convergence in $\mathscr{S}'$, we must show that, for each test function $\varphi \in \mathscr{S}$, the partial sums

$$\left\langle \sum_{n=-m}^{m} c_n \cdot \delta(t - n), \varphi(t) \right\rangle = \sum_{n=-m}^{m} c_n \cdot \varphi(n)$$

converge as $m \to \infty$, and that the limit $\in \mathscr{S}'$. Since the c_n are slowly increasing (dominated by a fixed power of $|n|$), whereas $\varphi(n)$ is rapidly decreasing (decreasing faster than any power of $|n|$), the convergence is trivial, as asserted.

The main part of Theorem 5.13 is a little harder. We know only that, for $\varphi \in \mathscr{D}$ (not $\varphi \in \mathscr{S}$),

$$\langle \hat{T}, \varphi \rangle = \sum_{n=-\infty}^{\infty} c_n \cdot \varphi(n),$$

where, since $\varphi \in \mathscr{D}$, the sum is finite. That is, we have no guarantee that the corresponding series for $\varphi \in \mathscr{S}$ even converges. We have to rule out the possibility that the definition of $\langle \hat{T}, \varphi \rangle$ for $\varphi \in \mathscr{S}$ might be given by some completely ersatz process.

The proof is topological, based on the fact that $\hat{T} \in \mathscr{S}'$, i.e. that $\hat{T}$ is a continuous linear functional on $\mathscr{S}$. Suppose the theorem is false. Then the coefficients c_n grow faster than the nth power of $|n|$ for any k; more precisely there is a subsequence $n_k = n(k)$ of indices n such that $|n(k)| \to \infty$, and

$$|c_{n(k)}| > |n(k)|^k \text{ for all } k.$$

Let θ be a test function with support on $[-\frac{1}{2}, \frac{1}{2}]$ and such that $\theta(0) = 1$, as in the last lemma above. Define

$$\theta_k(t) = \theta[t - n(k)] / |n(k)|^k.$$

Then $\theta_k \to 0$ in $\mathscr{S}$. For the size of $\theta[t - n(k)]$ and its derivatives is unaltered by translation (from $\theta(t)$ to $\theta[t - n(k)]$), whereas the coefficients $(1/|n(k)|^k)$ decrease faster than any fixed power of $|n(k)|$ as $k \to \infty$.

On the other hand, $|\langle \hat{T}, \theta_k \rangle| = |c_{n(k)} \cdot \theta_k[n(k)]| = |c_{n(k)} / |n(k)|^k| > 1$. Thus $\theta_k \to 0$ in $\mathscr{S}$, but $\langle \hat{T}, \theta_k \rangle$ does not approach zero, contradicting the continuity of $\hat{T}$.

Q.E.D.

Proof of Theorem 5.10, completed. So far, we have proved this theorem only for test function $\varphi \in \mathscr{D}$. However, once we have Theorem 5.13, the extension to $\varphi \in \mathscr{S}$ is trivial. (Here compare the trivial part of Theorem 5.13 itself.) That is, once we know that the coefficients c_n are slowly increasing, the extension of the formula

$$\langle \hat{T}, \varphi \rangle = \sum_{n=-\infty}^{\infty} c_n \cdot \varphi(n)$$

from $\varphi \in \mathscr{D}$ to $\varphi \in \mathscr{S}$ is immediate. (The extension gives a continuous functional on $\mathscr{S}$. And, of course, since $\mathscr{D}$ is dense in $\mathscr{S}$, the extension from $\mathscr{D}$ to $\mathscr{S}$ is unique.)

Remark. In this presentation, the topological steps have been isolated in Theorem 5.13. On the other hand, Theorem 5.13, which shows that a sequence $\{c_n\}$ occurs as the Fourier coefficients of a periodic distribution if and only if $\{c_n\}$ is slowly increasing, is of interest in its own right.

Computation of the Fourier coefficients. We still need to prove the Consistency Theorem 5.12 (which gives the traditional formula for the Fourier coefficients in the case where $T = T_f$ corresponds to a periodic function $f(x)$). This will follow from the next result, which applies to arbitrary periodic distributions.

Theorem 5.14. Let T be a tempered distribution which is periodic with period 1. Let $\varphi \in \mathscr{S}$ be any test function such that $\int_{-\infty}^{\infty} \varphi(x)\,dx = 1$. Then the nth Fourier coefficient c_n of T is given by

$$c_n = \lim_{a \to \infty} \langle e^{-2\pi inx}T(x), a^{-1}\varphi(x/a)\rangle.$$

Remark. Loosely speaking, we can regard this as 'integrating' $e^{-2\pi inx}T(x)$: precisely what we should expect. For, as $a \to \infty$, the mass of the function $a^{-1}\varphi(x/a)$ becomes spread out over many periods of T, while the dilated function $\varphi(x/a)$ becomes nearly constant on each interval of length 1; the constant a^{-1} maintains the total mass of $a^{-1}\varphi(x/a)$ at one.

Proof. Without loss of generality, we can take $n = 0$. For $[e^{-2\pi inx}T(x)]\hat{\ }(t) = \hat{T}(t + n)$, and the zeroth Fourier coefficient of $\hat{T}(t + n)$ (corresponding to $t = 0$) equals the nth Fourier coefficient for $\hat{T}$ and T. Now (with $n = 0$),

$$\langle T, a^{-1}\varphi(x/a)\rangle = \langle T\hat{\ }\check{\ }, a^{-1}\varphi(x/a)\rangle = \langle T\hat{\ }, \check{\varphi}(at)\rangle,$$

since $[a^{-1}\varphi(x/a)]\check{\ }(t) = \check{\varphi}(at)$. Furthermore, from the representation of $\hat{T}$ as a sequence of delta functions (Theorem 5.10),

$$\langle \hat{T}, \check{\varphi}(at)\rangle = \sum_{n=-\infty}^{\infty} c_n\check{\varphi}(an).$$

Now the c_n are slowly increasing (Theorem 5.13), whereas $\check{\varphi}(t)$ is rapidly decreasing (since $\check{\varphi} \in \mathscr{S}$). Hence, as $a \to \infty$, the contribution to $\sum c_n\check{\varphi}(an)$ of all terms with $n \neq 0$ becomes negligible; i.e.

$$\lim_{a \to \infty} \sum_{n=-\infty}^{\infty} c_n\check{\varphi}(an) = c_0\check{\varphi}(0).$$

Since $\int_{-\infty}^{\infty} \varphi(x)\,dx = \check{\varphi}(0) = 1$, this proves Theorem 5.14.

Proof of Theorem 5.12. We suppose that $T = T_f$, where $f(x)$ is periodic with period 1, and $f(x)$ is integrable over $[0, 1]$, and we want to prove

$$c_n = \int_0^1 e^{-2\pi i n x} f(x) \, dx.$$

As above, we can assume $n = 0$. This would follow immediately from Theorem 5.14 if we could take as our 'test function' the indicator function of $[0, 1]$:

$$\varphi(x) = \begin{cases} 1 & \text{for } 0 \leqslant x \leqslant 1 \\ 0 & \text{elsewhere.} \end{cases}$$

Unfortunately, this is not a test function. So we replace it by a mesa function φ_ε (see Figure 9) to which Theorem 5.14 applies. Then we let $\varepsilon \to 0$. Further details are left to the reader.

5 Further examples – Fourier series

We shall derive a batch of the standard Fourier series identities by arguments so 'soft' that they are practically scandalous. On the other hand, if the reader thinks it would be easier simply to compute the formulas, rather than go through all of this theory, he or she is undoubtedly right. We have not claimed that doing things the easy way is necessarily easy.

Example 1. Consider the distribution

$$T(x) = \sum_{n=-\infty}^{\infty} \delta(x - n).$$

This is periodic with period 1; hence, $\hat{T}(t)$ is a sum of delta functions:

$$\hat{T}(t) = \sum_{n=-\infty}^{\infty} c_n \cdot \delta(t - n).$$

Figure 9. A mesa function φ_ε.

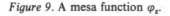

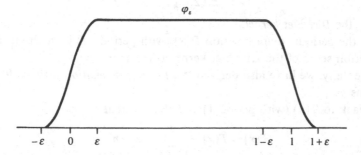

On the other hand, T itself is such a sum of delta functions, whence $\hat{T}$ is periodic with period 1. Hence, all of the coefficients c_n are equal, and

$$\hat{T}(t) = C \cdot \sum_{n=-\infty}^{\infty} \delta(t-n), \qquad C = \text{constant}.$$

Thus $\hat{T}$ is a constant multiple of T, and, by the Fourier Inversion Theorem,

$$T = T^{\wedge\vee} = C^2 \cdot T,$$

whence $C^2 = 1$, $C = \pm 1$.

To show that $C = +1$, we observe that there exist test functions $\varphi(x) > 0$ with $\hat{\varphi}(t) > 0$, e.g. $\varphi(x) = e^{-\pi x^2}$. Thus, $\langle T, \varphi \rangle > 0$ and $\langle T, \hat{\varphi} \rangle = \langle \hat{T}, \varphi \rangle = C \cdot \langle T, \varphi \rangle > 0$, whence $C > 0$.

Hence the periodic delta distribution T above is its own Fourier transform.

Now, since $T(x) = \sum \delta(x-n)$, and $[\delta(x-n)]^{\vee}(t) = e^{2\pi i n t}$, direct computation gives $T^{\vee}(t) = \sum e^{2\pi i n t}$. (Here, of course, convergence of the series is in the distribution sense – i.e. in the sense of $\mathscr{S}'$. Since $\sum \delta(x-n)$ converges in $\mathscr{S}'$, and the Fourier transform is continuous there, we conclude that $\sum e^{2\pi i n t}$ converges in $\mathscr{S}'$.) From the fact that $T^{\vee} = T$, we deduce the identity:

$$T(x) = \sum_{n=-\infty}^{\infty} \delta(x-n) = \sum_{n=-\infty}^{\infty} e^{2\pi i n x}.$$

In many applications, it is conventional to use period 2π instead of period 1. Setting

$$D(x) = (1/2\pi) \cdot T(x/2\pi),$$

and recalling that $\delta(x/2\pi) = 2\pi \cdot \delta(x)$, we have

$$D(x) = \sum_{n=-\infty}^{\infty} \delta(x - 2n\pi) = \frac{1}{2\pi} \cdot \sum_{n=-\infty}^{\infty} e^{inx},$$

or, in its more conventional form,

$$D(x) = \frac{1}{\pi}\left(\frac{1}{2} + \sum_{n=1}^{\infty} \cos nx\right).$$

The partial sum

$$D_n(x) = \frac{1}{\pi}\left(\frac{1}{2} + \sum_{k=1}^{n} \cos kx\right)$$

is called the *Dirichlet kernel*.

Thus the periodic delta function $D(x)$ (with period 2π) is the limit in the distribution sense of the Dirichlet kernel $D_n(x)$ as $n \to \infty$.

Incidentally, we have also derived the *Poisson summation formula* for test functions φ.

Go back to $T(x)$ (with period 1) and the fact that

$$T(x) = \hat{T}(x) = \sum_{n=-\infty}^{\infty} \delta(x-n).$$

Since, by definition, $\langle \hat{T}, \varphi \rangle = \langle T, \hat{\varphi} \rangle$, we have, for any test function φ,

$$\sum_{n=-\infty}^{\infty} \varphi(n) = \sum_{n=-\infty}^{\infty} \hat{\varphi}(n)$$

(where, as usual, $\hat{\varphi}(t) = \int_{-\infty}^{\infty} e^{-2\pi i t x} \varphi(x)\, dx$). This is the Poisson sum formula.

Example 2. Consider the classical formulas

$$2 \cdot \sum_{n=1}^{\infty} \frac{\sin nx}{n} = \begin{cases} \pi - x, & 0 < x \leqslant \pi \\ -\pi - x, & -\pi \leqslant x < 0 \end{cases}$$

and

$$\frac{4}{\pi} \sum_{n \, \text{odd}} \frac{\sin nx}{n} = \begin{cases} 1, & 0 < x < \pi \\ -1, & -\pi < x < 0. \end{cases}$$

Of course, these are easily verified by computing the Fourier coefficients in the traditional way. However, we regard that approach as 'mindless'. We intend to derive these formulas from symmetry considerations, as in Example 1 above.

Begin with the first formula. Let $f(x)$ be the function defined there, i.e. $f(x) = \pi - x$ for $0 < x \leqslant \pi$, $f(x) = -\pi - x$ for $-\pi \leqslant x < 0$, and extend f to be periodic with period 2π. Then,

$$f'(x) = -1 + 2\pi \cdot \sum_{n=-\infty}^{\infty} \delta(x - 2\pi n)$$

$$= -1 + 2\pi \cdot D(x) \quad \text{(cf. Example 1)}$$

$$= -1 + 2 \cdot \left(\frac{1}{2} + \sum_{n=1}^{\infty} \cos nx \right)$$

(again using Example 1), whence

$$f(x) = 2 \cdot \sum_{n=1}^{\infty} \frac{\sin nx}{n} + C, \quad C = \text{constant}.$$

Since f is an odd function, $C = 0$.

Now consider the second formula. Again following Example 1, we have

$$D(x) - D(x - \pi) = \frac{1}{\pi} \sum_{n=1}^{\infty} [\cos nx - \cos n(x - \pi)]$$

$$= \frac{2}{\pi} \sum_{n \, \text{odd}} \cos nx.$$

On the other hand, the periodic function $g(x)$, defined on the right hand side of the second formula, satisfied

$$g'(x) = 2[D(x) - D(x - \pi)].$$

Integrating, and observing that g is also an odd function, we have

$$g(x) = \frac{4}{\pi} \sum_{n \, \text{odd}} \frac{\sin nx}{n}.$$

(In fact, most of the standard Fourier series identities – involving piecewise linear functions, segments of parabolas and the like – can be obtained in this way. One starts with the 2π-periodic delta function $D(x)$, translates it, and integrates. These integrations, applied to the formula

$$D(x) = \frac{1}{\pi}\left(\frac{1}{2} + \sum_{n=1}^{\infty} \cos nx\right)$$

can be read off at a glance: no hard computations of integrals are required. There is one exception – for functions which are not odd, the constant of integration ($=$ the zeroth Fourier coefficient) must be computed directly.)

Example 3. If we differentiate these formulas instead of integrating them, we obtain results which are valid in distribution theory, but which have no classical analogs. For instance, differentiating the formula for $D(x)$ in Example 1, we obtain

$$\sum_{n=-\infty}^{\infty} \delta'(x - 2\pi n) = \frac{-1}{\pi} \cdot \sum_{n=1}^{\infty} n \cdot \sin nx.$$

Example 4. For the convenience of the reader, we shall extend the general formulas for periodic distributions from the case of period 1 to period T.

Note. 'T' is the traditional letter for periods. Thus we shall use it, and for obvious reasons in this example we shall denote a general periodic distribution by S.

Let S be a tempered distribution with period T, i.e. $S(x + T) = S(x)$. Then $S(xT)$ has period 1. Since $[S(xT)]\hat{} = T^{-1}\hat{S}(t/T)$, the key formulas of Section 4 become

$$T^{-1}\hat{S}(t/T) = \sum_{n=-\infty}^{\infty} c_n \cdot \delta(t - n),$$

$$\hat{S}(t) = T \cdot \sum_{n=-\infty}^{\infty} c_n \cdot \delta(Tt - n),$$

and since $\delta(Tu) = T^{-1}\delta(u)$, this leads to the final formula:

$$\hat{S}(t) = \sum_{n=-\infty}^{\infty} c_n \cdot \delta\left(t - \frac{n}{T}\right).$$

Thus, $\hat{S}$ involves a sequence of delta functions located at the points 0, $\pm 1/T$, $\pm 2/T$, ...
From the Fourier Inversion Formula $S = S\hat{}\check{}$, we obtain

$$S(x) = \sum_{n=-\infty}^{\infty} c_n \cdot e^{2\pi inx/T}.$$

Finally, for the case where $S = S_f$ comes from a function $f(x)$ with period T, the formula $c_n = \int_0^1 e^{-2\pi inx} f(xT)\, dx$ becomes

$$c_n = \frac{1}{T}\int_0^T e^{-2\pi inx/T} f(x)\, dx.$$

6

Extension to higher dimensions

1 Multi-indices

In this chapter we extend the theory from $\mathbb{R}^1$ to $\mathbb{R}^q$ = the space of real q-tuples $(x_1, \ldots, x_q)$; vectors in $\mathbb{R}^q$ will be written in boldface; thus $\mathbf{x} = (x_1, \ldots, x_q)$ denotes a typical vector in $\mathbb{R}^q$.

The extension from $\mathbb{R}^1$ to $\mathbb{R}^q$ involves no really new ideas. It does bring in, however, a proliferation of subscripts and other notational clutter. For example, in $\mathbb{R}^1$ there is only one derivative operator d/dx, and the most general higher order derivative is simply $(d/dx)^n$. By contrast, in $\mathbb{R}^q$ there are q first order derivatives $\partial/\partial x_i$, $1 \leqslant i \leqslant q$, leading to the rather elaborate higher order expression

$$\frac{\partial^{\alpha_1 + \alpha_2 + \cdots \alpha_q}}{\partial x_1^{\alpha_1} \, \partial x_2^{\alpha_2} \cdots \partial x_q^{\alpha_q}}. \tag{1}$$

Obviously one wants an abbreviation for (1). The standard form uses *multi-indices*, which are defined as q-tuples of non-negative integers,

$$\alpha = (\alpha_1, \ldots, \alpha_q).$$

We define

$$|\alpha| = \alpha_1 + \alpha_2 + \cdots + \alpha_q.$$

Then the differential operator in (1) is abbreviated as:

$$\mathbf{D}^\alpha = \partial^{|\alpha|}/\partial \mathbf{x}^\alpha.$$

We will at times express $\mathbf{D}^\alpha f$ as $f^{(\alpha)}$. Similarly, when dealing with polynomials in q variables, we use the abbreviation

$$\mathbf{x}^\alpha = x_1^{\alpha_1} x_2^{\alpha_2} \cdots x_q^{\alpha_q}.$$

For use with Taylor's formula, we set

$$\alpha! = \alpha_1! \cdot \alpha_2! \cdots \alpha_q!.$$

Then the Taylor series of an analytic function of q variables is abbreviated as

$$f(\mathbf{x}) = \sum_\alpha \frac{1}{\alpha!} \mathbf{D}^\alpha f(\mathbf{a})(\mathbf{x} - \mathbf{a})^\alpha.$$

Examples. Let $q = 3$, and, dropping the subscripts in this example, replace (x_1, x_2, x_3) by the standard notation for three-dimensional space (x, y, z). Consider $f(x, y, z)$, and let

$$\alpha = (1, 0, 2).$$

Then

$$D^\alpha f = \frac{\partial^{|\alpha|}}{\partial x^\alpha} f = \frac{\partial^3}{\partial x\, \partial z^2} f(x, y, z).$$

The corresponding term in the Taylor series of f about $(0, 0, 0)$ is

$$\frac{1}{\alpha!} (D^\alpha f)(0) \cdot x^\alpha = \frac{1}{1! \cdot 0! \cdot 2!} \left[\frac{\partial^3 f}{\partial x\, \partial z^2} (0, 0, 0) \right] xz^2.$$

While we are dealing with notations, there are a few more conventions concerning $\mathbb{R}^q$ which we may as well set down. Translation of a function f on $\mathbb{R}^q$ by a constant vector $\mathbf{a}$ is given by

$$f(\mathbf{x} - \mathbf{a}) = f(x_1 - a_1, \ldots, x_q - a_q).$$

The dilatation of vectors $\mathbf{x}$ by a non-zero scalar factor a (positive or negative) is given by

$$a\mathbf{x} = (ax_1, \ldots, ax_q).$$

More generally, let A be any non-singular real $(q \times q)$ matrix. Then we may consider the linear transformation of $\mathbb{R}^q$ onto itself given by

$$\mathbf{y} = A\mathbf{x}.$$

Here the vectors would, of course, be displayed in column form:

$$\begin{pmatrix} y_1 \\ \vdots \\ y_q \end{pmatrix} = A \begin{pmatrix} x_1 \\ \vdots \\ x_q \end{pmatrix}.$$

(We adopt the convention that a point $\mathbf{x}$ in $\mathbb{R}^q$ may be written either as a row vector or a column vector, according to context. Lest this seem shocking, we observe that there is no need to regard matrix notation with religious awe. A point in $\mathbb{R}^q$ is really an ordered q-tuple of numbers, and remains the same thing no matter how one writes it.)

As is standard, we use the notation $\mathbf{x} \cdot \mathbf{y}$ for the inner or 'dot' product of two q-vectors:

$$\mathbf{x} \cdot \mathbf{y} = x_1 y_1 + x_2 y_2 + \cdots + x_q y_q.$$

We define the norm $\|\mathbf{x}\|$ by

$$\|\mathbf{x}\| = (\mathbf{x} \cdot \mathbf{x})^{\frac{1}{2}} = (x_1^2 + \cdots + x_q^2)^{\frac{1}{2}}.$$

(In some arguments it is convenient to use the 'sup' norm $\|\mathbf{x}\|_\infty = \max\{|x_i| : 1 \leqslant i \leqslant q\}$. For a fixed dimension q, this is equivalent in view of the fact that

$$\|\mathbf{x}\|_\infty \leqslant \|\mathbf{x}\| \leqslant q^{\frac{1}{2}} \|\mathbf{x}\|_\infty.)$$

In connection with multiple integration, we write

$$\int_{\mathbb{R}^q} f(\mathbf{x})\, d\mathbf{x} = \underbrace{\iiint \cdots \int}_{(q\ \text{times})} f(x_1, \ldots, x_q)\, dx_1\, dx_2 \cdots dx_q.$$

That is, $d\mathbf{x}$ denotes integration with respect to the q-dimensional 'volume' (or 'content') in $\mathbb{R}^q$.

2 Fourier transforms on $\mathbb{R}^q$

Here we must extend the contents of Chapter 4 (the classical Fourier transform) from $\mathbb{R}^1$ to $\mathbb{R}^q$. We emphasize that these results – like those in Chapter 4 – deal with ordinary functions and have nothing *per se* to do with distributions. Of course, they form the prerequisite for the distribution-theoretic treatment which follows.

There are several approaches which one could follow. One, which has a certain appeal, is simply to assert: 'The extension from $\mathbb{R}^1$ to $\mathbb{R}^q$ is trivial!' Then we state the Fourier Inversion Theorem for $\mathbb{R}^q$, as given at the end of this section, and go on. Readers who prefer this approach can achieve it merely by skipping this section.

However, the extension of the Fourier transform to $\mathbb{R}^q$ actually does require a little work – hardly profound, but not entirely trivial either. We thought we had better set it down.

First we extend the notation of an open support test function from $\mathbb{R}^1$ to $\mathbb{R}^q$.

An *open support test function* $\varphi(x_1, \ldots, x_q)$ on $\mathbb{R}^q$ is a C^∞ function such that every partial derivative $\varphi^{(\alpha)}$ is *rapidly decreasing*: i.e. for every integer N, the product $\|\mathbf{x}\|^N \cdot \varphi^{(\alpha)}(\mathbf{x})$ remains bounded as $\|\mathbf{x}\| \to \infty$. The set of all open support test functions on $\mathbb{R}^q$ is denoted by $\mathscr{S}(\mathbb{R}^q)$.

As in Chapter 4, we first develop the Fourier transform for the very restricted class $\mathscr{S}(\mathbb{R}^q)$. Then in the next section we give the development which parallels Chapter 5: tempered distributions on $\mathbb{R}^q$.

Our approach is as follows. We will make use of the results which we already have (for $\mathbb{R}^1$), and show how they can be extended to $\mathbb{R}^q$. Thus the most difficult work has already been done (in Chapter 4). Mercifully, we do *not* need to begin anew and redo the whole theory in order to arrive at the q-dimensional case. The key to this approach is contained in the following.

Definition. Let $\varphi(x_1, \ldots, x_q) \in \mathscr{S}(\mathbb{R}^q)$ and let k be an integer, $1 \leqslant k \leqslant q$. Then the *$k$th partial Fourier transform* and *kth partial inverse Fourier transform* of φ, written $F_k(\varphi)$ and $F_k^{-1}(\varphi)$, respectively, are defined by

$$F_k(\varphi) = \int_{-\infty}^{\infty} e^{-2\pi i t_k x_k} \varphi(x_1, \ldots, x_k, \ldots, x_q)\, dx_k;$$

$$F_k^{-1}(\varphi) = \int_{-\infty}^{\infty} e^{2\pi i t_k x_k} \varphi(x_1, \ldots, x_k, \ldots, x_q)\,dx_k.$$

Thus, $F_k(\varphi)$ and $F_k^{-1}(\varphi)$ are functions of variables $x_1, \ldots, x_{k-1}, t_k, x_{k+1}, \ldots, x_q$.

Notice that we are holding all the variables but x_k fixed. Therefore, $\varphi(x_1, \ldots, x_k, \ldots, x_q)$ works as a function of x_k alone and F_k and F_k^{-1} are the one-dimensional Fourier and inverse Fourier transform of φ. Hence, by applying the one-dimensional Fourier Inversion Theorem, we have the following lemma.

Lemma. $F_k(F_k^{-1}(\varphi)) = F_k^{-1}(F_k(\varphi)) = \varphi$, for all $\varphi \in \mathscr{S}(\mathbb{R}^q)$.

Proof. $\varphi(x_1, \ldots, x_q)$ with all but one variable fixed is an open support test function on $\mathbb{R}^1$. Hence, by the Fourier Inversion Theorem in Chapter 4, we are done.

Lemma. For any $\varphi \in \mathscr{S}(\mathbb{R}^q)$, its kth partial Fourier transform $F_k(\varphi)$ and inverse Fourier transform $F_k^{-1}(\varphi)$ are in $\mathscr{S}(\mathbb{R}^q)$.

Proof. We will consider $k = 1$ and, for brevity, write $(x_2, \ldots, x_q) = \mathbf{y}$. Thus, $\varphi(x_1, x_2, \ldots, x_q) = \varphi(x_1, \mathbf{y})$ and $F_1(\varphi)(t_1, x_2, \ldots, x_q) = F_1(\varphi)(t_1, \mathbf{y})$.

$F_1(\varphi)$ is a C^∞ function because $e^{-2\pi i t_1 x_1}\varphi(x_1, \mathbf{y})$ is C^∞ and we can differentiate under the integral sign as many times as we wish.

We first show that $F_1(\varphi)$ is bounded. (Then boundedness, together with the Identities (b) and (c) in Chapter 4, will show that $F_1(\varphi) \in \mathscr{S}(\mathbb{R}^q)$.) Since $\varphi(x_1, \mathbf{y})$ is rapidly decreasing, $|\varphi(x_1, \mathbf{y})| \leqslant \text{Const}/(1 + x_1^2)$ for some constant, and therefore

$$|F_1(\varphi)| \leqslant \int_{-\infty}^{\infty} |e^{-2\pi i t_1 x_1}| \cdot |\varphi(x_1, \mathbf{y})|\,dx_1 \leqslant \int_{-\infty}^{\infty} \text{Const}/(1 + x_1^2)\,dx_1 < M.$$

Hence, $F_1(\varphi)$ is bounded independently of $(t_1, \mathbf{y})$.

Now we must show that $F_1(\varphi) \in \mathscr{S}(\mathbb{R}^q)$, i.e. that $F_1(\varphi)$ is rapidly decreasing together with all of its derivatives. Here we shall first consider $F_1(\varphi)$ itself and ignore the derivatives. Then we extend the result to include the derivatives.

To show that $F_1(\varphi)$ is rapidly decreasing, we must show that 'bounded' implies 'rapidly decreasing'. If we can show that $t_1^m \cdot \|\mathbf{y}\|^{2n} \cdot F_1(\varphi)$ is bounded for all m and n, it will follow that $F_1(\varphi)$ is rapidly decreasing. Now,

$$t_1^m \cdot \|\mathbf{y}\|^{2n} \cdot F_1(\varphi) = t_1^m \int_{-\infty}^{\infty} e^{-2\pi i t_1 x_1} \|\mathbf{y}\|^{2n} \varphi(x_1, \mathbf{y})\,dx_1$$

$$= t_1^m F_1(\|\mathbf{y}\|^{2n} \cdot \varphi(x_1, \mathbf{y}))$$

$$= \text{Const} \cdot F_1\left[\left(\frac{d}{dx_1}\right)^m \|\mathbf{y}\|^{2n} \varphi(x_1, \mathbf{y})\right] \quad \text{(by Identity (b) in Chapter 4).}$$

Since $\|y\|^{2n} = (\sqrt{(x_2^2 + \cdots + x_q^2)})^{2n} = (x_2^2 + \cdots + x_q^2)^n$ is a polynomial, $\|y\|^{2n} \cdot \varphi(x_1, y)$ is in $\mathscr{S}(\mathbb{R}^q) \cdot$ (!). Hence, so is its mth derivative $(d/dx_1)^m$. As we have already seen, this implies the boundedness of $F_1[(d/dx_1)^m \|y\|^{2n} \varphi(x_1, y)]$, as desired. Therefore, $F_1(\varphi)$ is rapidly decreasing.

For the partial derivatives of $F_1(\varphi)$, there are two cases: $\partial/\partial x_1$ and $\partial/\partial x_i$ with $i > 1$. In the first case, the partial derivative and the partial Fourier transform both involve the same variable x_1; in the second case they involve different variables. To handle the first case, we use Identities (b) and (c) of Chapter 4. The result then follows in the same fashion as above. The second case is trivial: for when we have different variables x_1 and x_i, the x_1 partial Fourier transform and the partial derivative $\partial/\partial x_i$ commute.

The inverse transform $F_k^{-1}(\varphi)$ is, of course, handled in the same way.

We now define the Fourier transform on $\mathbb{R}^q$.

Definition. Let $\varphi(x_1, \ldots, x_q) \in \mathscr{S}(\mathbb{R}^q)$. We define the Fourier transform $\hat{\varphi}$ and the inverse Fourier transform $\check{\varphi}$ of φ as follows:

$$\hat{\varphi}(t_1, \ldots, t_q) = \underbrace{\int_{-\infty}^{\infty} \cdots \int_{-\infty}^{\infty}}_{(q \text{ times})} e^{-2\pi i(t_1 x_1 + \cdots + t_q x_q)} \varphi(x_1, \ldots, x_q)\, dx_1 \cdots dx_q;$$

$$\check{\varphi}(t_1, \ldots, t_q) = \underbrace{\int_{-\infty}^{\infty} \cdots \int_{-\infty}^{\infty}}_{(q \text{ times})} e^{2\pi i(t_1 x_1 + \cdots + t_q x_q)} \varphi(x_1, \ldots, x_q)\, dx_1 \cdots dx_q;$$

We may rewrite the Fourier transform as an iteration of the partial Fourier transforms defined above. That is, we take the partial transform with respect to x_1, then x_2, then x_3, ... up to x_q.

Lemma. Recalling that F_k denotes the partial Fourier transform with respect to x_k, we have

$$\hat{\varphi}(t_1, \ldots, t_q) = F_q(F_{q-1}(\cdots(F_1(\varphi))\cdots);$$
$$\check{\varphi}(t_1, \ldots, t_q) = F_q^{-1}(F_{q-1}^{-1}(\cdots(F_1^{-1}(\varphi))\cdots).$$

Proof. This is just the replacement of a q-dimensional (volume) integral by an iterated integral.

Corollary. If $\varphi \in \mathscr{S}(\mathbb{R}^q)$, then $\hat{\varphi} \in \mathscr{S}(\mathbb{R}^q)$ and $\check{\varphi} \in \mathscr{S}(\mathbb{R}^q)$.

Proof. This was proved above for the partial Fourier transform. Since the q-dimensional transform is an iteration of partial transforms, the result follows.

Since φ is integrable in $\mathbb{R}^q$, we can interchange the order of integration, which gives the following lemma.

Lemma.

$$\phi = F_{k_1}(F_{k_2}\cdots(F_{k_q}(\varphi))\cdots)$$

and

$$\check{\phi} = F_{k_1}^{-1}(F_{k_2}^{-1}\cdots(F_{k_q}^{-1}(\varphi))\cdots),$$

for any rearrangement $(k_1, \ldots, k_q)$ of $(1, \ldots, q)$.

Now we can prove the Fourier Inversion Theorem on $\mathbb{R}^q$.

Theorem (Fourier Inversion Theorem). For any $\varphi(x_1, \ldots, x_q)$ in $\mathscr{S}(\mathbb{R}^q)$, $\varphi^{\hat{}\vee} = \varphi^{\vee\hat{}} = \varphi$.

Proof. By definition and by the above lemma,

$$\varphi^{\hat{}\vee} = F_1^{-1}(F_2^{-1}\cdots(F_q^{-1}(\phi)\cdots)$$
$$= F_1^{-1}(F_2^{-1}\cdots(F_q^{-1}(F_q(F_{q-1}\cdots(F_1(\varphi)\cdots).$$

Let $\psi_q(t_1, \ldots, t_{q-1}, x_q) = F_{q-1}(F_{q-2}\cdots(F_1(\varphi)\cdots)$. Then by the above lemmas, $\psi_q \in \mathscr{S}(\mathbb{R}^q)$ and $F_q^{-1}(F_q(\psi_q)) = \psi_q$. Thus F_q^{-1} and F_q collapse to an identity. Repeating this q times, we have $\varphi^{\hat{}\vee} = \varphi$. Likewise, $\varphi^{\vee\hat{}} = \varphi$.   Q.E.D.

In what follows, we are going to extend Chapter 2 (general distributions) and Chapter 5 (tempered distributions) from $\mathbb{R}^1$ to $\mathbb{R}^q$. Tempered distributions on $\mathbb{R}^q$ will get priority in our treatment because we will be working with them in the next chapter. Since the motivations have been amply discussed in earlier chapters, we will take a 'definition–theorem–proof' viewpoint for the most part. Furthermore, some of the proofs, which are virtually identical to those in earlier chapters, will be either sketched or omitted.

3 Tempered distributions on $\mathbb{R}^q$

For coherence, and since the last section was described as optional, we repeat the following:

Definition. An *open support test function* $\varphi(\mathbf{x})$ on $\mathbb{R}^q$ is a function such that:
  (a) $\varphi(\mathbf{x})$ is C^∞ (continuously differentiable with respect to any partial derivative D^α);
  (b) every partial derivative of φ is rapidly decreasing (i.e., for every integer N and multi-index α, the product $\|\mathbf{x}\|^N \cdot \varphi^{(\alpha)}(\mathbf{x})$ remains bounded as $\|\mathbf{x}\| \to \infty$).

Definition. A sequence of open support test functions $\varphi_n(\mathbf{x})$ *converges to zero* (write $\varphi_n \to 0$) if for each pair consisting of an integer N and a multi-index

α, the sequence of functions $\|\mathbf{x}\|^N \cdot \varphi_n^{(\alpha)}(\mathbf{x})$ approaches zero uniformly on $\mathbb{R}^q$ as $n \to \infty$.

We say that $\varphi_n \to \varphi$ if $(\varphi - \varphi_n) \to 0$.

Definition. A *tempered distribution* T is a mapping from the set of open support test functions into the complex numbers that satisfies the following:
  (a) (Linearity.) $\langle T, a\varphi(\mathbf{x}) + b\psi(\mathbf{x}) \rangle = a \cdot \langle T, \varphi(\mathbf{x}) \rangle + b\langle T, \psi(\mathbf{x}) \rangle$, for all test functions φ, ψ and all complex constants a, b;
  (b) (Continuity.) If $\varphi_n \to 0$ in the sense defined above, then $\langle T, \varphi_n \rangle \to 0$.

The set of all open support test functions is denoted by $\mathscr{S}$ or $\mathscr{S}(\mathbb{R}^q)$ and the set of all tempered distributions is denoted by $\mathscr{S}'$ or $\mathscr{S}'(\mathbb{R}^q)$. As in Chapter 5, we will use the term 'convergence in $\mathscr{S}$' for the convergence defined above.

Remark. In the one-dimensional theory of Chapters 1 to 5, we have mostly avoided any mention of measure theory. We could continue on this path: distribution theory is logically independent of measure theory; but our exposition would suffer from a lack of good examples. Namely, in q dimensions, there is no analog of the class of piecewise continuous functions which can be discussed intelligently without some notion of the 'content' or 'measure' of a q-dimensional region. Thus, in our examples, we shall make free use of the term 'measurable function'. Readers who do not know measure theory will get on perfectly well if they simply interpret the term 'measurable function' to mean 'any function for which integration makes sense'.

A measurable function is called *slowly increasing* if there is some integer N and some constant such that:
$$|f(\mathbf{x})| \leqslant \text{Const} \cdot (1 + \|\mathbf{x}\|)^N.$$
This has the effect that the integral
$$\int_{\mathbb{R}_q} f(\mathbf{x})\varphi(\mathbf{x}) \, d\mathbf{x}$$
exists for all open support test functions φ. (Recall that 'dx' denotes $dx_1 \cdots dx_q$; i.e. the integration is with respect to volume in $\mathbb{R}^q$.) In what follows, we shall assume that all of the functions which we use are measurable, so that 'slowly increasing' means 'slowly increasing and measurable'.

Definition. Let $f(\mathbf{x})$ be a slowly increasing function on $\mathbb{R}^q$. Then we define the distribution T_f corresponding to f by
$$\langle T_f, \varphi \rangle = \int_{\mathbb{R}_q} f(\mathbf{x})\varphi(\mathbf{x}) \, d\mathbf{x},$$
for all $\varphi \in \mathscr{S}(\mathbb{R}^q)$.

Definition (Calculus operations). Let S and T be tempered distributions. Then we define new distributions $S + T, cT$ (for complex constants c), $(\partial/\partial x_j)T(\mathbf{x})$, $T(A\mathbf{x})$ (where A is a non-singular $(q \times q)$ matrix), $T(\mathbf{x} - \mathbf{a})$ (for a constant vector $\mathbf{a}$) and $g(\mathbf{x})T(\mathbf{x})$ (where $g(\mathbf{x})$ is a C^∞ function which is slowly increasing together with all of its derivatives) by:

(1) $\langle S + T, \varphi \rangle = \langle S, \varphi \rangle + \langle T, \varphi \rangle$;

(2) $\langle cT, \varphi \rangle = c \cdot \langle T, \varphi \rangle$;

(3) $\langle (\partial/\partial x_j)T(\mathbf{x}), \varphi \rangle = - \langle T, (\partial/\partial x_j)\varphi(\mathbf{x}) \rangle$;

(4) $\langle T(A\mathbf{x}), \varphi \rangle = |\det A|^{-1} \langle T, \varphi(A^{-1}\mathbf{x}) \rangle$, where $\det A$ is the determinant of A and A^{-1} is the inverse matrix of A;

(5) $\langle T(\mathbf{x} - \mathbf{a}), \varphi \rangle = \langle T, \varphi(\mathbf{x} + \mathbf{a}) \rangle$;

(6) $\langle g(\mathbf{x})T(\mathbf{x}), \varphi \rangle = \langle T, g(\mathbf{x})\varphi(\mathbf{x}) \rangle$.

Definition (Fourier transform). The Fourier transform $\hat{T}$ and inverse Fourier transform $\check{T}$ of a tempered distribution T are defined by:

(7) $\langle \hat{T}, \varphi \rangle = \langle T, \hat{\varphi} \rangle$;

(7a) $\langle \check{T}, \varphi \rangle = \langle T, \check{\varphi} \rangle$.

We note that, just as for $\mathbb{R}^1$, $\check{T}(\mathbf{x}) = \hat{T}(-\mathbf{x})$.

Theorem 6.1. All of the operations in the preceding definitions regarding tempered distributions are tempered distributions.

Proof. Since condition (a) (linearity) is trivial, we will turn to condition (b) (continuity). We give the proof for $T(A\mathbf{x})$; the proofs for the others are similar to those in the one-dimensional case.

Recall that $\langle T(A\mathbf{x}), \varphi \rangle = |\det A|^{-1} \langle T, \varphi(A^{-1}\mathbf{x}) \rangle$ and that T itself is a distribution. Hence, to prove that $T(A\mathbf{x})$ is continuous, it suffices to show that $\varphi_n(\mathbf{x}) \to 0$ in $\mathscr{S}$ implies $\varphi_n(A^{-1}\mathbf{x}) \to 0$ in $\mathscr{S}$. Let $\varphi_n \to 0$ in $\mathscr{S}$ and let us show that $\|\mathbf{x}\|^N \cdot [\varphi_n(A^{-1}\mathbf{x})]^{(\alpha)} \to 0$ uniformly for each pair of an integer N and a multi-index α. By using an identity from calculus, we have

$$[\varphi_n(A^{-1}\mathbf{x})]^{(\alpha)} = \sum_{|\beta| \leq |\alpha|} \text{Const}_\beta \cdot \varphi_n^{(\beta)}(A^{-1}\mathbf{x}),$$

where Const_β is a constant that depends on β. Now, since $\varphi_n \to 0$ in $\mathscr{S}$, $\|\mathbf{x}\|^N \varphi_n^{(\beta)}(A^{-1}\mathbf{x}) \to 0$ uniformly for each β. Hence, their finite linear combination $\|\mathbf{x}\|^N [\varphi_n(A^{-1}\mathbf{x})]^{(\alpha)} \to 0$ uniformly on $\mathbb{R}^q$, as desired.

Remark. In the above, we defined

$$\langle (\partial/\partial x_i)T, \varphi \rangle = - \langle T, (\partial/\partial x_i)\varphi(\mathbf{x}) \rangle.$$

By applying the definition repeatedly, we may define any mixed partial derivative of T. Thus

$$(\partial^{|\alpha|}/\partial x_{i_1}^{\alpha_{i_1}} \cdots \partial x_{i_q}^{\alpha_{i_q}})T = \partial^{\alpha_{i_1}}/\partial x_{i_1}^{\alpha_{i_1}}(\partial^{\alpha_{i_2}}/\partial x_{i_2}^{\alpha_{i_2}} \cdots (\partial^{\alpha_{i_q}}/\partial x_{i_q}^{\alpha_{i_q}}T) \cdots),$$

where $x_{i_1}, \ldots, x_{i_q}$ is any rearrangement of the variables $x_1, \ldots, x_q$.

An important fact about distributions in several variables is that partial derivatives commute. For example, in $\mathbb{R}^2$, where we write the variables as x, y for clarity, we have

$$\frac{\partial^2 T}{\partial x\, \partial y} = \frac{\partial^2 T}{\partial y\, \partial x}$$

for any distribution T. This extends, *mutatis mutandis*, to higher dimensions and higher order mixed partial derivatives. The reason why it holds is that, in the basic definition above, the derivative operator is passed from the distribution over to the test function. The test functions, being C^∞, allow the partial derivatives to commute without restriction.

The fact that the identity $\partial^2 T/\partial x\, \partial y = \partial^2 T/\partial y\, \partial x$ holds without restriction is one of the many attractive features of distribution theory. The example below shows that partial derivatives do not always commute in classical analysis.

Example. Take $q=2$ and write (x_1, x_2) as (x, y). Let

$$f(x, y) = \begin{cases} xy \cdot \dfrac{x^2 - y^2}{x^2 + y^2} & \text{if } (x, y) \neq (0, 0) \\ 0 & \text{if } (x, y) = (0, 0). \end{cases}$$

Then $\partial^2 f/\partial x\, \partial y$ and $\partial^2 f/\partial y\, \partial x$ exist for all $(x, y) \in \mathbb{R}^2$ and are continuous except at $(x, y) = (0, 0)$. Nevertheless, $(\partial^2/\partial x\, \partial y)f(0, 0) = 1$ and $(\partial^2/\partial y\, \partial x)f(0, 0) = -1$.

(Of course, if we view $f(x, y)$ as a distribution, rather than as a function, then the partial derivatives commute.)

As an obvious extension of the above we have:

Theorem 6.2 (Mixed partial derivatives). Let T be a distribution. Then any mixed partial derivative of T commutes. That is

$$(\partial^{|\alpha|}/\partial x_{i_1}^{\alpha_{i_1}}\, \partial x_{i_2}^{\alpha_{i_2}} \cdots \partial x_{i_q}^{\alpha_{i_q}})T = (\partial^{|\alpha|}/\partial x_1^{\alpha_1}\, \partial x_2^{\alpha_2} \cdots \partial x_q^{\alpha_q})T,$$

for any rearrangement $i_1, \ldots, i_q$ of $1, 2, \ldots, q$.

(Although the above theorem could be combined with the other identity theorems which follow, we stated it separately since it has no analog for one-dimensional distributions.)

Theorem 6.3 (Consistency Theorem). Let $f(x)$ and $e(x)$ be slowly increasing functions. Further restrictions are given in parts (3) and (7) below. Then:

(1) $T_f + T_e = T_{f(x) + e(x)}$;

(2) $c \cdot T_f = T_{cf(\mathbf{x})}$ (c = complex constant);

(3) $\partial/\partial x_j (T_f) = T_{[\partial/\partial x_j f(\mathbf{x})]}$, where f is continuous and $\partial/\partial x_j f(\mathbf{x})$ is slowly increasing and continuous with respect to the variable x_j;

(4) $(T_f)(A\mathbf{x}) = T_{f(A\mathbf{x})}$ (A is a non-singular matrix);

(5) $(T_f)(\mathbf{x} - \mathbf{a}) = T_{f(\mathbf{a}-\mathbf{x})}$ (a is a constant vector);

(6) $g(\mathbf{x})(T_f) = T_{g(\mathbf{x})f(\mathbf{x})}$, where $g(\mathbf{x})$ is C^∞ and slowly increasing together with all of its derivatives;

(7) $(T_f)\hat{} = T_{\hat{f}}$, where f is integrable on $\mathbb{R}^q$.

Proof. We give proofs for (3) and (4); the others are left to the reader. Consider $\partial/\partial x_j(T_f)$. By definition,

$$\langle \partial/\partial x_j(T_f), \varphi \rangle = -\langle T_f, \partial/\partial x_j \varphi(\mathbf{x}) \rangle = -\int_{\mathbb{R}^q} f(\mathbf{x})[\partial/\partial x_j \varphi(\mathbf{x})]\, d\mathbf{x}.$$

Since f is slowly increasing and $\partial/\partial x_j \varphi(\mathbf{x})$ is rapidly decreasing, their product is rapidly decreasing. This guarantees convergence of the integrals involved. Now we use the Fubini Theorem to interchange the order of integration:

$$-\int_{\mathbb{R}^{q-1}} \int_{-\infty}^{\infty} f(\mathbf{x})[\partial/\partial x_j \varphi(\mathbf{x})]\, dx_j\, d\mathring{\mathbf{x}},$$

where $\mathring{\mathbf{x}}$ is a $(q-1)$-tuple vector $(x_1, \ldots, x_{j-1}, x_{j+1}, \ldots, x_q)$.

Now we integrate the integral with respect to dx_j by parts and use the fact that $\varphi(\mathbf{x}) \to 0$ as $x_j \to \pm\infty$ to get

$$\int_{\mathbb{R}^{q-1}} \int_{-\infty}^{\infty} [\partial/\partial x_j f(\mathbf{x})]\varphi(\mathbf{x})\, dx_j\, d\mathring{\mathbf{x}}.$$

Applying the Fubini Theorem once more, we have

$$\int_{\mathbb{R}^q} [\partial/\partial x_j f(\mathbf{x})]\varphi(\mathbf{x})\, d\mathbf{x} = \langle T_{\partial/\partial x_j f(\mathbf{x})}, \varphi \rangle.$$

Consider $T_f(A\mathbf{x})$. By definition,

$$\langle T_f(A\mathbf{x}), \varphi \rangle = |\det A|^{-1}\langle T_f, \varphi(A^{-1}\mathbf{x}) \rangle = |\det A|^{-1}\int_{\mathbb{R}^q} f(\mathbf{x})\varphi(A^{-1}\mathbf{x})\, d\mathbf{x}.$$

Make the substitution $\mathbf{u} = A^{-1}\mathbf{x}$. Then by the Jacobian formula of multi-variable calculus, $d\mathbf{u} = |\text{Jacobian of } \mathbf{u} \text{ with respect to } \mathbf{x}|\, d\mathbf{x} = |\det A^{-1}|\, d\mathbf{x}$. Thus we obtain

$$\int_{\mathbb{R}^q} f(A\mathbf{u})\varphi(\mathbf{u})\, d\mathbf{u} = \langle T_{f(A\mathbf{x})}, \varphi \rangle.$$

One word about the absolute value in the Jacobian formula. Suppose φ and f are both positive. Then the integral $\int f(\mathbf{x})\varphi(A^{-1}\mathbf{x})\, d\mathbf{x}$, both before the change of variable and after, must be positive. To obtain that we need the absolute value of the Jacobian. The point is that here the integrals are of functions over subsets of $\mathbb{R}^q$, and we associate no direction or orientation to these subsets.

Definition. Let $\{T_n\}$ be a sequence of tempered distributions, and let T be a tempered distribution. We say that T_n converges to T, written $T_n \to T$, if
$$\langle T_n, \varphi \rangle \to \langle T, \varphi \rangle \text{ for every open support test function } \varphi.$$

Theorem 6.4 (Continuity of the distribution operations). Let $\{S_n\}$ and $\{T_n\}$ be sequences of tempered distributions converging to tempered distributions S and T, respectively, and let the constants be the same as in the definition of calculus operations. Then:

(1) $(S_n + T_n) \to S + T$;

(2) $c \cdot T_n \to c \cdot T$;

(3) $\partial/\partial x_j T_n(\mathbf{x}) \to \partial/\partial x_j T(\mathbf{x})$;

(4) $T_n(A\mathbf{x}) \to T(A\mathbf{x})$;

(5) $T_n(\mathbf{x} - \mathbf{a}) \to T(\mathbf{x} - \mathbf{a})$;

(6) $g(\mathbf{x})T_n(\mathbf{x}) \to g(\mathbf{x})T(\mathbf{x})$;

(7) $\hat{T}_n \to \hat{T}$.

Proof. The proofs are virtually identical to those in the one-dimensional case.

Theorem 6.5 (Identities of calculus). Let S and T be tempered distributions, $g(\mathbf{x})$ a C^∞ function slowly increasing together with its derivatives, A a non-singular $(q \times q)$ matrix, $\mathbf{a}$ a constant vector and $c \neq 0$ a complex constant. Then:

(1) $\partial/\partial x_j (S + T) = \partial/\partial x_j S + \partial/\partial x_j T$;

(2) $\partial/\partial x_j (c \cdot T) = c \cdot (\partial/\partial x_j T)$;

(3) $\partial/\partial x_j [T(A\mathbf{x})] = \left(\sum_{i=1}^{q} a_{ij} \cdot \partial/\partial x_i T \right)(A\mathbf{x})$, where $A = (a_{ij})$;

(4) $\partial/\partial x_j [T(\mathbf{x} - \mathbf{a})] = (\partial/\partial x_j T)(\mathbf{x} - \mathbf{a})$;

(5) $\partial/\partial x_j [g(\mathbf{x})T(\mathbf{x})] = [\partial/\partial x_j g(\mathbf{x})]T(\mathbf{x}) + g(\mathbf{x})[\partial/\partial x_j T(\mathbf{x})]$.

Proof. We will prove (3). The rest of the proofs are similar to those in the one-dimensional case.

By definition,
$$\langle \partial/\partial x_j [T(A\mathbf{x})], \varphi \rangle = -\langle T(A\mathbf{x}), \partial/\partial x_j \varphi(\mathbf{x}) \rangle$$
$$= -|\det A|^{-1} \langle T(\mathbf{x}), (\partial/\partial x_j \varphi)(A^{-1}\mathbf{x}) \rangle.$$

On the other hand,
$$\left\langle \left(\sum_{i=1}^{q} a_{ij} \cdot \partial/\partial x_i T \right)(A\mathbf{x}), \varphi \right\rangle$$
$$= |\det A|^{-1} \left\langle \left(\sum_{i=1}^{q} a_{ij} \cdot \partial/\partial x_i T \right)(\mathbf{x}), \varphi(A^{-1}\mathbf{x}) \right\rangle$$
$$= -|\det A|^{-1} \left\langle T(\mathbf{x}), \left(\sum_{i=1}^{q} a_{ij} \cdot \partial/\partial x_i \right) \varphi(A^{-1}\mathbf{x}) \right\rangle$$
$$= -|\det A|^{-1} \left\langle T(\mathbf{x}), \left[\sum_{i=1}^{q} a_{ij} \left(\sum_{k=1}^{q} b_{ki} \cdot \partial/\partial x_k \varphi \right) \right](A^{-1}\mathbf{x}) \right\rangle,$$

where $(b_{ki}) = A^{-1}$. Now

$$\sum_{i=1}^{q} a_{ij} \left(\sum_{k=1}^{q} b_{ki} \cdot \partial/\partial x_k \varphi \right) = \sum_{k=1}^{q} \left(\sum_{i=1}^{q} a_{ij} b_{ki} \right) \partial/\partial x_k \varphi$$

and since

$$(b_{ki}) = A^{-1}, \ \sum_{i=1}^{q} a_{ij} b_{ki} = 0 \text{ if } k \neq j$$

and

$$\sum_{i=1}^{q} a_{ij} b_{ki} = 1 \text{ if } k = j.$$

Hence,

$$\left[\sum_{i=1}^{q} a_{ij} \left(\sum_{k=1}^{q} b_{ki} \cdot \partial/\partial x_k \varphi \right) \right] (A^{-1}\mathbf{x}) = (\partial/\partial x_j \varphi)(A^{-1}\mathbf{x}).$$

Thus

$$\left\langle \left(\sum_{i=1}^{q} a_{ij} \cdot \partial/\partial x_i T \right)(A\mathbf{x}), \varphi \right\rangle = -|\det A|^{-1} \langle T, (\partial/\partial x_j \varphi)(A^{-1}\mathbf{x}) \rangle.$$

Theorem 6.6 (Fourier transform identities). Let S, T and the constants be the same as in the Calculus Identities Theorem. Then:

(1) $(aT + bS)\hat{} = a\hat{T} + b\hat{S}$;

(2) $[\partial/\partial x_j T(\mathbf{x})]\hat{} = 2\pi i \cdot t_j \cdot \hat{T}(\mathbf{t})$;

(3) $[x_j T(\mathbf{x})]\hat{} = \dfrac{-1}{2\pi i} \cdot \partial/\partial t_j \hat{T}(\mathbf{t})$;

(4) $[T(\mathbf{x} - \mathbf{a})]\hat{} = e^{-2\pi i \mathbf{a} \cdot \mathbf{t}} \hat{T}(\mathbf{t})$;

(5) $[e^{2\pi i \mathbf{a} \cdot \mathbf{x}} T(\mathbf{x})]\hat{} = \hat{T}(\mathbf{t} - \mathbf{a})$;

(6) $[T(A\mathbf{x})]\hat{} = |\det A|^{-1} \cdot \hat{T}(A^{-1}\mathbf{t})$;

(7) $T\hat{\check{}} = T\check{\hat{}} = T$.

The proofs of these identities are so similar to proofs already given in Chapter 5 that we omit them.

Example (Dirac Delta Function). As in the one-dimensional case, we define the Dirac Delta Function on $\mathbb{R}^q$ as

$$\langle \delta(\mathbf{x}), \varphi(\mathbf{x}) \rangle = \varphi(0)$$

$$\langle \delta(\mathbf{x} - \mathbf{a}), \varphi(\mathbf{x}) \rangle = \varphi(\mathbf{a})$$

for all test functions φ on $\mathbb{R}^q$.

Sometimes we consider a 'partial delta function', which, instead of being concentrated in a point, corresponds to a lower dimensional subspace of $\mathbb{R}^q$. We illustrate this by an example. Let $q = 2$, and write the vector $\mathbf{x} \in \mathbb{R}^2$ as (x, y). Then the ordinary delta function would be written $\delta(x, y)$. We define the partial delta function $\delta(x)$ as a distribution concentrated over the subspace $x = 0$, i.e. along the y-axis:

$$\langle \delta(x), \varphi(x, y) \rangle = \int_{\mathbb{R}^1} \varphi(0, y) \, dy.$$

Loosely speaking, we could view $\delta(x)$ as the product of 'a delta function in x times the constant function 1 in y'. The presence of the constant $1 = 1(y)$ means that we integrate dy. Likewise

$$\langle \delta(y), \varphi(x, y) \rangle = \int_{\mathbb{R}^1} \varphi(x, 0) \, dx.$$

Clearly $\delta(x)$ and $\delta(y)$ are continuous and hence are distributions.

The partial delta functions give an example of a tensor product, a notion which will play a key role in the next chapter.

4 General distributions on $\mathbb{R}^q$

The following section is so obviously repetitive that we compress it drastically. No proofs will be given.

Definition. A function $\varphi(\mathbf{x})$ on $\mathbb{R}^q$ is called a (compact support) *test function* if:
  (a) $\varphi(\mathbf{x})$ is C^∞ (continuously differentiable with respect to any partial derivative D^α);
  (b) $\varphi(\mathbf{x})$ has compact support (i.e. $\varphi(\mathbf{x})$ vanishes outside of some compact ball $\{\mathbf{x}: \|\mathbf{x}\| \leqslant a\}$).

Definition. A sequence $\varphi_n(\mathbf{x})$ of compact support test functions *converges to zero* (write $\varphi_n \to 0$) if:
  (a) for each multi-index α the sequence of partial derivatives $D^\alpha \varphi_1(\mathbf{x})$, $D^\alpha \varphi_2(\mathbf{x}), \ldots$ converges uniformly to zero;
  (b) the φ_n have uniformly bounded support, i.e. there is a compact ball $K = \{\mathbf{x}: \|\mathbf{x}\| \leqslant a\}$, independent of n, such that every $\varphi_n(\mathbf{x})$ vanishes outside of K.
  We say that $\varphi_n \to \varphi$ if $(\varphi - \varphi_n) \to 0$.

Definition. A (general) *distribution* T is a mapping from a set of all compact support test functions into the real or complex numbers that satisfies the following:
  (a) (Linearity.) $\langle T, a\varphi + b\psi \rangle = a \langle T, \varphi \rangle + b \langle T, \psi \rangle$;
  (b) (Continuity.) If $\varphi_n \to 0$ in the sense defined above, then $\langle T, \varphi_n \rangle \to 0$.

We shall denote the set of all compact support test functions by $\mathscr{D}(\mathbb{R}^q)$ and the set of all general distributions by $\mathscr{D}'(\mathbb{R}^q)$.

A measurable function f is locally integrable if $\int_K |f(\mathbf{x})|\, d\mathbf{x} < \infty$ for any compact set K in $\mathbb{R}^q$.

Definition. Let $f(\mathbf{x})$ be a locally integrable function on $\mathbb{R}^q$. Then we define the distribution T_f corresponding to f by

$$\langle T_f, \varphi \rangle = \int_{\mathbb{R}^q} f(\mathbf{x})\varphi(\mathbf{x})\, d\mathbf{x}.$$

Definition. Let S and T be arbitrary general distributions. Then we define new distributions $S + T$, aT ($a = $ constant), $(\partial/\partial x_j)T(\mathbf{x})$, $T(A\mathbf{x})$ (A is a $q \times q$ non-singular matrix), $T(\mathbf{x} - \mathbf{a})$, and $g(\mathbf{x})T(\mathbf{x})$ (where $g(\mathbf{x})$ is a C^∞ function) by:
   (1) $\langle S + T, \varphi \rangle = \langle S, \varphi \rangle + \langle T, \varphi \rangle$;
   (2) $\langle aT, \varphi \rangle = a\langle T, \varphi \rangle$;
   (3) $\langle \partial/\partial x_j T(\mathbf{x}), \varphi \rangle = -\langle T, \partial/\partial x_j \varphi(\mathbf{x}) \rangle$;
   (4) $\langle T(A\mathbf{x}), \varphi \rangle = |\det A|^{-1}\langle T, \varphi(A^{-1}\mathbf{x}) \rangle$, where $\det A$ is the determinant of A and A^{-1} is the inverse matrix of A;
   (5) $\langle T(\mathbf{x} - \mathbf{a}), \varphi \rangle = \langle T, \varphi(\mathbf{x} + \mathbf{a}) \rangle$;
   (6) $\langle g(\mathbf{x})T(\mathbf{x}), \varphi \rangle = \langle T, g(\mathbf{x})\varphi(\mathbf{x}) \rangle$.

Theorem 6.7. All of the operators defined above are distributions.

(As stated above, in this subsection we omit the proofs.)

Definition. Let $\{T_n\}$ be a sequence of distributions and let T be a distribution. We say that T_n converges to T, written $T_n \to T$, if

$$\langle T_n, \varphi \rangle \to \langle T, \varphi \rangle$$

for every test function φ.

Consistency Theorem 6.8. Under hypothesis similar to those in Chapter 2, when an operation on distributions is well defined, both classically and in terms of distributions, then the two definitions coincide.

Continuity Theorem 6.9. All of the operations in Theorem 6.7 are continuous on $\mathscr{D}'(\mathbb{R}^q)$.

Theorem 6.10 (Identities Theorem). Let S and T be distributions, $g(\mathbf{x})$ a C^∞ function and A a non-singular $q \times q$ matrix. Then
   (1) $\partial/\partial x_j(S + T) = \partial/\partial x_j S + \partial/\partial x_j T$;
   (2) $\partial/\partial x_j(aT) = a(\partial/\partial x_j T)$, where $a = $ constant;
   (3) $\partial/\partial x_j[T(A\mathbf{x})] = \sum_{i=1}^q (a_{ij}\, \partial/\partial x_i T)(A\mathbf{x})$, where a_{ij} is the (i, j)th component of A;

(4) $\partial/\partial x_j[T(\mathbf{x} - \mathbf{a})] = (\partial/\partial x_j T)(\mathbf{x} - \mathbf{a})$;

(5) $\partial/\partial x_j[g(\mathbf{x})T(\mathbf{x})] = [\partial/\partial x_j g(\mathbf{x})] \cdot T(\mathbf{x}) + g(\mathbf{x})[\partial/\partial x_j T(\mathbf{x})]$;

(6) $\partial^2/\partial x_i\, \partial x_j(T) = \partial^2/\partial x_j\, \partial x_i(T)$.

To conclude this chapter, we repeat that the theory of distributions in q-dimensions is an obvious extension of the one-dimensional theory. The only really new result is the fact that mixed partial derivatives commute: e.g. $\partial^2 T/\partial x\, \partial y = \partial^2 T/\partial y\, \partial x$. Of course, many mathematical theories which use distributions – e.g. differential equations – are far more difficult in higher dimensions than in one dimension. But the theory of distributions itself is not.

Besides the obvious facts that many problems in physics and analysis are q-dimensional, there is another reason for introducing higher dimensional distributions at this stage. Certain important problems in distribution theory – even in the one-dimensional case – are best solved by moving up to higher dimensions and then coming back down again. This is the subject of the next chapter.

7

A general definition of multiplication and convolution for distributions

1 Introduction

The definitions of multiplication and convolution which we have used so far, while adequate for many purposes, are subject to severe limitations. Basically they are unsymmetrical. Thus, for multiplication, we have defined the product gT only for certain C^∞ functions g, although we allow complete freedom for the distribution T. Similarly for the convolution $S * T$, we required T to have compact support, while allowing complete freedom for S.

For many purposes in analysis, one needs symmetrical definitions. Thus the product of two continuous functions $g(x)h(x)$ is well defined, whether the first function g is C^∞ or not. In a sense, the 'good' qualities of h (being continuous, rather than a general distribution) balance the 'bad' qualities of g (not being C^∞). The point is that such a simple and important case as this – the product of two continuous functions – is *not* defined within the traditional theory of distributions.

There are two ways out of this difficulty. The first (and by far the most often practiced) is simply to acknowledge that distribution theory is a partial theory – sometimes useful and sometimes awkward. Accordingly, if one wants to multiply two continuous functions g and h, one passes *out* of distribution theory into classical analysis, multiplies g and h in the standard way within classical analysis, and then passes back *into* distribution theory. This works perfectly well, but from the viewpoint of distribution theory it seems slightly unfortunate.

The second approach is to expand the distribution-theoretic definitions of multiplication and convolution to cover the cases routinely encountered in analysis. That is what we will do in this chapter.

Here we state a few cautions. We will not give a 'universal' definition of multiplication and convolution for distributions, i.e. one which allows the multiplication/convolution of arbitrary distributions. There is strong evidence that such a universal definition does not exist. Rather, we give a *symmetrical* definition, which is far more general than those given previously (in Chapters 1–6), and which covers the most important cases.

The material in this chapter is a little more difficult than that covered so far in this deliberately elementary book. But it is coherent with the theme of the book, which is to treat distribution theory as an extension of classical analysis, rather than as an application of high-powered theories. For the reader who wants to skim, the time-honored tactic of reading the theorems and skipping the proofs will make the basic ideas easy to see.

There has been considerable research in this area, of which we mention König [KO], Shiraishi [SR], and Horváth [HV] as representative sources. However, so far as we know, very little of this research has found its way into the textbook literature.

Finally we mention the thought, and the hope, that non-standard analysis (a theory which allows the use of 'infinitesimal quantities') may some day provide a much simpler treatment of these questions. However, that must lie outside the scope of this book.

Notations and standing conventions

Throughout this chapter, unless stated otherwise, all distributions will be tempered distributions. Similarly, 'test function' will mean open support test function.

In Chapter 7, unlike Chapter 6, we will not represent vectors $x \in \mathbb{R}^q$ in boldface. This inconsistency conforms to standard usage at the two (distinct) levels at which Chapters 6 and 7 are written. At the advanced calculus level (Chapter 6) it is standard to use boldface for vectors. At the graduate level (Chapter 7) it is standard not to. We are simply following those conventions.

By *complementary subspaces* U and V of $\mathbb{R}^q$ we mean the following. We are given subspaces U, V of $\mathbb{R}^q$, of dimensions r and s, respectively, with $r + s = q$ and such that

$$U \perp V \quad (U \text{ is orthogonal to } V),$$
$$U + V = \mathbb{R}^q.$$

That is, we require that $u \in U$, $v \in V$ implies $u \cdot v = 0$, and that each vector $x \in \mathbb{R}^q$ has a decomposition of the form $x = u + v$, $u \in U$, $v \in V$. Then, by the orthogonality condition, this decomposition is unique.

We assume that the subspaces U and V are provided with orthonormal bases $e_1, \ldots, e_r$ and $f_1, \ldots, f_s$, respectively. That is,

$$e_i \cdot e_j = \begin{cases} 1 & \text{for } i = j \\ 0 & \text{for } i \neq j, \end{cases}$$

and similarly for the f_i.

In terms of these bases, we have coordinate systems for U and V, respectively. Thus for U: if u is a vector in U, and

$$u = u_1 e_1 + \cdots + u_r e_r,$$

then we say that $\{u_1, \ldots, u_r\}$ are the *coordinates* of u. (Of course, $u_1, \ldots, u_r$ are scalars, whereas $e_1, \ldots, e_r$ and u are vectors.)

Similarly, we define the *coordinates* $\{v_1, \ldots, v_s\}$ of any vector $v \in V$ by the equation

$$v = v_1 f_1 + \cdots + v_s f_s.$$

In integration formulas we write

$$du = du_1 \cdots du_r \quad \text{and} \quad dv = dv_1 \cdots dv_s.$$

Because the bases $e_1, \ldots, e_r$ and $f_1, \ldots, f_s$ for U and V are orthonormal, we have

$$du \, dv = dx = dx_1 \cdots dx_q;$$

that is, $du \, dv$ gives the correct q-dimensional measure on $\mathbb{R}^q$.

We recall that the Fourier transform on $\mathbb{R}^q$ has the form

$$\hat{\phi}(t) = \int \cdots \int_{\mathbb{R}^q} e^{-2\pi i t \cdot x} \varphi(x) \, dx,$$

where

$$t \cdot x = t_1 x_1 + \cdots + t_q x_q.$$

Note that both $t, x \in \mathbb{R}^q$, but that we have used two different letters for the sake of clarity.

We need similar notations for the Fourier transform in terms of u and v, but somehow the introduction of four different letters becomes cumbersome. So we write

$$\hat{u} = \text{'the variable corresponding to } t\text{' for } u,$$

$$\hat{v} = \text{'the variable corresponding to } t\text{' for } v.$$

Sometimes, but not always, we use

$$\hat{x} = t.$$

Again, since the bases $e_1, \ldots, e_r$ and $f_1, \ldots, f_s$ are orthonormal, the inner product comes out the same way, whether computed in terms of t and x, or in terms of $\hat{u}, u, \hat{v}$ and v. That is, if

$$t = \hat{u} + \hat{v} \quad \text{and} \quad x = u + v,$$

then

$$t \cdot x = \hat{u} \cdot u + \hat{v} \cdot v = \hat{u}_1 u_1 + \cdots + \hat{u}_r u_r + \hat{v}_1 v_1 + \cdots + \hat{v}_s v_s.$$

Hence, in terms of $\hat{u}$, u, $\hat{v}$ and v, the Fourier transform can be written

$$\hat{\phi}(\hat{u}, \hat{v}) = \int \cdots \int_{\mathbb{R}^q = U + V} e^{-2\pi i (\hat{u} \cdot u + \hat{v} \cdot v)} \varphi(u, v) \, du \, dv.$$

Often it is convenient to reverse the roles of u and $\hat{u}$ (and likewise v and $\hat{v}$), and thus write

$$\phi(u, v) = \int \cdots \int_{\mathbb{R}^q = U + V} e^{-2\pi i(u \cdot \hat{u} + v \cdot \hat{v})} \varphi(\hat{u}, \hat{v}) \, d\hat{u} \, d\hat{v}.$$

Clearly, both of the above displayed formulas for ϕ say the same thing, and it is merely a matter of convenience which one we use.

2 Distributions depending on a parameter

Let $\mathbb{R}^q$ be decomposed into complementary subspaces, $\mathbb{R}^q = U + V$, as described at the end of the introduction above. We recall that the spaces U and V are orthogonal, of dimensions r and s, respectively, with $r + s = q$.

We shall consider tempered distributions $T_v(u)$ on U, depending on a parameter $v \in V$. (Thus the distribution $T_v(u)$ is r-dimensional, and the parameter v is s-dimensional.) Sometimes the family $\{T_v(u)\}$ can be combined to give a single tempered distribution $T(x) = T(u, v)$ on $\mathbb{R}^q$. Then we will call the distributions $T_v(u)$, $v = v_o = $ constant, the *cross sections* of the global distribution $T(x)$.

It should be emphasized that not every distribution $T(x)$ can be broken down into cross sections $T_v(u)$. In this way, of course, distributions differ from ordinary functions. With an ordinary function, we could simply set $v = v_o$, and replace the function $T(x) = T(u, v)$ on $\mathbb{R}^q$ by the cross section $T(u, v_o)$, defined on U. However, distributions, unlike functions, are not defined pointwise – they are defined by their action on test functions. It turns out that some distributions $T(x)$ on $\mathbb{R}^q$ have cross sections, and others do not. The cases where they do, however, are important enough to deserve consideration.

Examples. Let $q = 2$, and use the conventional (x, y) variables for $\mathbb{R}^2$. Let U be the x-axis, and let V be the y-axis.

(a) Let $T(x, y)$ be the two-dimensional delta function $\delta(x, y)$. That is, for each test function φ on $\mathbb{R}^2$,

$$\langle \delta, \varphi \rangle = \varphi(0, 0).$$

Then T cannot be broken down into cross sections $T_y(x)$ along the lines $\{y = y_o = $ constant$\}$ parallel to the x-axis. We shall not give the impossibility proof here. But the intuitive idea is easy to see. If there were such a decomposition, then $T_y(x)$ would be zero for $y \neq 0$, and $T_0(x)$ would have an infinite positive mass situated at $x = 0$. It can be proved that an infinite positive mass situated on a compact set (with no negative mass to cancel it) does not exist in distribution theory.

(b) Again let $q = 2$, and let U, V be as above. However, here let δ be the one-dimensional delta function, and set $T(x, y) = \delta(x - y)$. More precisely, T maps each test function $\varphi(x, y)$ into its integral along the line $\{y = x\}$:

$$\langle T, \varphi \rangle = \int_{-\infty}^{\infty} \varphi(y, y) \, dy.$$

This time T does have cross sections, and these cross sections are just what we would expect: for each fixed y,

$$T_y(x) = \delta(x - y).$$

The reader can easily check that this special case fulfils all of the conditions set down in the general definition below.

Definition. Let $T(x) = T(u, v)$ be a tempered distribution on $\mathbb{R}^q = U + V$, and let $\{T_v(u)\}$ be a family of tempered distributions on U, parametrized by $v \in V$. We say that the $T_v(u)$ form a family of *cross sections* for $T(x)$ if the following two conditions are satisfied:

(1) (Reconstitution of $T(x)$ from the cross sections $T_v(u)$.) For each open support test function $\psi(u, v)$,

$$\langle T(x), \psi(u, v) \rangle = \int_V \langle T_v(u), \psi(u, v) \rangle \, dv.$$

(2) (Strong continuity.) If $v \to v_0$ in V and $\varphi_n \to \varphi$ in $\mathscr{S}$, then
$$\langle T_v(u), \varphi_n(u) \rangle \to \langle T_{v_0}(u), \varphi(u) \rangle.$$

Remarks. We observe that the integral in (1) above is an ordinary integral, of the scalar quantity $\langle T_v(u), \psi(u, v) \rangle$ (obtained by holding v fixed). This scalar quantity is a function of v, and in writing down the integral we have implicitly assumed that this function is integrable.

The notion of continuity used in (2) is stronger than the 'weak continuity' which has been mostly used in this book. Weak continuity would only require that for any fixed test function φ (and not for a sequence $\{\varphi_n\}$, $\varphi_n \to \varphi$), $\langle T_v, \varphi \rangle \to \langle T_{v_0}, \varphi \rangle$. We have introduced this stronger notion of continuity here for the first time because this was the first time we needed it.

In Section 4 we will introduce a notion, *localization*, which is more general than the *cross sections* considered here. However, as so often happens, 'more general' does not mean 'easier to deal with'. The cross sections, when they do exist, are far more intuitive and easier to work with than the more general localizations. Furthermore, cross sections do exist in many important cases. This will be illustrated in Section 6.

3 Tensor products

Tensor products play an important role in the higher theory of distributions. We have not needed them so far in this book. Now we do.

To focus the ideas, we emphasize at the outset that the tensor product corresponds, in classical analysis, to a very special type of product – a product involving *independent* variables. Thus consider $\mathbb{R}^2$ with its usual x, y coordinate system. The product of x^2 with $\sin y$ to produce $x^2 \sin y$ is a tensor product. The product of x^2 with x^2 to produce x^4 is not. More generally, the tensor product corresponds, in classical analysis, to products of the form $f(x)g(y)$, with independent variables x and y.

As we shall see, the tensor product of two distributions always exists. By contrast, more general products (which, unlike tensor products, may involve a repetition of the same variable) do not always exist. For example, it can be shown that the square of the Dirac Delta Function, $\delta(x)^2$, has no distribution-theoretic meaning.

One main point of this chapter is to determine in general when the product of two distributions does exist. The tensor product – a special case – will serve as a tool for attacking the general problem.

We continue to use the notations for complementary subspaces U and V of $\mathbb{R}^q$ set in the introduction to this chapter.

Definition. Let U and V be complementary subspaces of $\mathbb{R}^q$, and S and T be tempered distributions on U and V, respectively. Then the *tensor product* $S(u)T(v)$ is defined, for any open support test function $\varphi(u, v) \in \mathscr{S}(\mathbb{R}^q)$, by the formula

$$\langle S(u)T(v), \varphi(u, v) \rangle = \langle S(u), \langle T(v), \varphi(u, v) \rangle \rangle.$$

Notes. For any fixed $u \in U$, $\varphi(u, v)$ is a test function on V, and the function $\psi(u) = \langle T(v), \varphi(u, v) \rangle$ is, as will be shown in the Appendix, a test function on U. Hence the above tensor product $S(u)T(v)$ is well defined.

In many texts, the tensor product is written $S(u) \otimes T(v)$. We have preferred the notation $S(u)T(v)$ in accordance with the overall plan of this book, which is to follow the notations of classical analysis whenever possible. Indeed, as we shall see, when $S = S_f$ and $T = T_g$ correspond to ordinary functions $f(u)$ and $g(v)$, then $S(u)T(v)$ corresponds to the product $f(u)g(v)$.

On the other hand, we must remember that a general definition of 'product' for distributions has not – so far – been given. Indeed, the above is our first step in this direction. It is restricted, however, to the case where u and v are 'independent' variables – i.e. variables belonging to complementary subspaces U, V as above.

Theorem A. The tensor product $S(u)T(v)$ is a tempered distribution on $\mathbb{R}^q$.

Theorem B. The tensor product is commutative, i.e. $S(u)T(v) = T(v)S(u)$.

The proofs of Theorems A and B are sketched in Appendix 3. We have two reasons for putting them aside: they are (a) rather long, and (2) not very enlightening. By contrast, those arguments which have to do with the actual *operations* of analysis – e.g. Fourier transforms, multiplication, convolution – will be brought front and center.

Corollary B1 (Commutation with derivatives and with integrals). Let $\varphi(u, v)$ be a test function in $\mathscr{S}(U + V)$ and $T(v)$ be a tempered distribution on V. Then

$$(\partial^{|\alpha|}/\partial u^\alpha)\langle T(v), \varphi(u, v)\rangle = \langle T(v), (\partial^{|\alpha|}/\partial u^\alpha)\varphi(u, v)\rangle;$$

$$\int_U \langle T(v), \varphi(u, v)\rangle \, du = \left\langle T(v), \int_U \varphi(u, v) \, du \right\rangle.$$

Proof of corollary. Let δ be the Dirac Delta Function on the subspace U. We recall that, for any test function $\varphi(u)$ on U, and any point $a \in U$, $(\partial^{|\alpha|}/\partial u^\alpha)\varphi(a) = (-1)^{|\alpha|}\langle(\partial^{|\alpha|}/\partial u^\alpha)\,\delta(u-a), \varphi\rangle$. Thus the first formula above follows from Theorem B with $S(u) = (-1)^{|\alpha|}(\partial^{|\alpha|}/\partial u^\alpha)\,\delta(u-a)$.

Similarly, the second formula above follows from Theorem B where $S(u)$ is the constant function 1 on the subspace U. For, given any test function $\varphi(u)$ on U, $\langle 1, \varphi \rangle = \int_U \varphi(u) \, du$.

We recall from the 'Notations' subsection at the end of Section 1 that, in Fourier transform formulas, we write $[T(x)]\hat{\ }$ as $\hat{T}(t)$, and use the variables $\hat{u}, \hat{v}$, with $\hat{u} \in U$ and $\hat{v} \in V$, so that $x = u + v$ but $t = \hat{u} + \hat{v}$.

Observe that we do not need to define the Fourier transform of the tensor product $[S(u)T(v)]\hat{\ }$. For $S(u)T(v)$ is a tempered distribution on $\mathbb{R}^q$, and hence its Fourier transform has already been defined in Chapter 6. We do, however, have the following basic identity.

Theorem 7.1. Let $S(u)$ and $T(v)$ be tempered distributions on U and V, respectively. Then

$$[S(u)T(v)]\hat{\ } = \hat{S}(u)\hat{T}(v).$$

Proof. Let F_u and F_v denote the partial Fourier transforms with respect to the u and v variables, respectively. That is, $F_v(\varphi(u, v)) = \int_V e^{-2\pi i(\hat{v} \cdot v)}\varphi(u, \hat{v}) \, d\hat{v}$, and similarly for F_u. (Here compare the partial Fourier transforms used in Chapter 6.) Let $\varphi(u, v) \in \mathscr{S}(U + V)$ and consider:

$$\langle [S(u)T(v)]\hat{\ }, \varphi(u, v)\rangle = \langle S(u)T(v), \hat{\varphi}(u, v)\rangle$$

$$= \langle S(u), \langle T(v), \hat{\varphi}(u, v)\rangle\rangle$$

$$= \langle S(u), \langle T(v), F_u[F_v(\varphi(u, v))]\rangle\rangle.$$

We have that $F_u[F_v(\varphi(u, v))] = \int_U e^{-2\pi i(\hat{u} \cdot u)}F_v(\varphi(\hat{u}, v)) \, d\hat{u}$. Furthermore $e^{-2\pi i(\hat{u} \cdot u)}F_v(\varphi(\hat{u}, v))$ is an open support test function. Now we wish to apply the distribution $T(v)$ to this integral and then bring the distribution inside the integral to obtain $\langle T(v), F_u[F_v(\varphi(u, v))]\rangle = F_u\langle T(v), F_v(\varphi(u, v))\rangle$. This is valid by Corollary B1. Hence, the last above displayed formula becomes

$$\langle S(u), F_u\langle T(v), F_v(\varphi(u, v))\rangle\rangle = \langle S(u), F_u\langle \hat{T}(v), \varphi(u, v)\rangle\rangle$$
$$= \langle \hat{S}(u)\hat{T}(v), \varphi(u, v)\rangle$$

as desired.

Theorem 7.2 (Consistency Theorem). Let $f(u)$ and $g(v)$ be slowly increasing functions on U and V, respectively, and let $S_f(u)$ and $T_g(v)$ be the corresponding distributions. Then the tensor product $S_f(u)T_g(v)$ is the distribution corresponding to the ordinary product $f(u)g(v)$ on $\mathbb{R}^q$.

Proof. We simply compute. The absolute convergence of the integrals is guaranteed by the fact that f and g are slowly increasing, whereas the test functions are rapidly decreasing. Now, by definition of the tensor product,

$$\langle S_f(u)T_g(v), \varphi(u, v)\rangle = \langle S_f(u), \langle T_g(v), \varphi(u, v)\rangle\rangle$$
$$= \int_U f(u) \int_V g(v)\varphi(u, v)\,dv\,du.$$

By Fubini's Theorem, this becomes

$$\int_{U+V} f(u)g(v)\varphi(u, v)\,dv\,du = \langle T_{f(u)g(v)}, \varphi(u, v)\rangle,$$

as desired.

4 Localization and partial integration

As in Sections 2 and 3, we deal with tempered distributions on $\mathbb{R}^q$, and we assume that $\mathbb{R}^q$ is decomposed into the direct sum of two orthogonal complementary subspaces, $\mathbb{R}^q = U + V$. For a tempered distribution $T(x) = T(u, v)$ on $\mathbb{R}^q$, our objective is to define

(1) the *localization* (or 'restriction') $T(u, 0)$ of T to the subspace U;
(2) the *partial integral* $S(u) = \int_V T(u, v)\,dv$.

This task is made more difficult by the fact that the localization or partial integral does not always exist! Nevertheless, the cases where they do exist are important in applications, and hence deserve consideration. Moreover, these notions provide the key to the definition of multiplication or convolution for distributions, as treated in Section 5.

Here we shall follow the plan previously used in Chapters 1 and 2. That is (following Chapter 1), we shall first find out what the defining formulas *ought* to be. Then (as in Chapter 2), we turn these formulas into definitions. However, here we can save space by applying a certain compression. The

Chapter 1-style formulas which we develop first actually serve as the 'Consistency Theorems' for the general definitions which come later. We shall label them as such. Thus, exactly as in Chapters 2 and 5, the Consistency Theorems show that the general definitions – when applied to the classical cases – give the same results as we would obtain if we did the calculations in the classical manner. Putting the Consistency Theorems first simply allows us to discover the general definitions naturally, instead of producing them *deux ex machina*.

Here we recall that a function f on $\mathbb{R}^q$ is said to be 'L^1' if f is measurable and $\int_{\mathbb{R}^q} |f| \, dx < \infty$.

Consistency Theorem 7.3a. Let f be L^1 on $\mathbb{R}^q$. Then

$$\int_{\mathbb{R}^q} \cdots \int f(x) \, dx = \hat{f}(0).$$

Proof. We simply recall that

$$\hat{f}(t) = \int_{\mathbb{R}^q} \cdots \int e^{-2\pi i t \cdot x} f(x) \, dx,$$

and observe that if $t = (t_1, \ldots, t_q) = (0, \ldots, 0)$, then $e^{-2\pi i t \cdot x} = 1$.

Consistency Theorem 7.3b. Let $f(x) = f(u, v)$ be L^1 on $\mathbb{R}^q$, and consider the partial integral

$$g(u) = \int_V f(u_1, \ldots, u_r, v_1, \ldots, v_s) \, dv_1 \cdots dv_s$$

$$= \int_V f(u, v) \, dv.$$

Take an open support test function $\varphi(u)$ on U, and consider the action $\langle g, \varphi \rangle$ of g on φ. Then

$$\langle g, \varphi \rangle = \int_{\mathbb{R}^q} \cdots \int f(u, v) \varphi(u) \, dv \, dv$$

$$= \int_{\mathbb{R}^q} \cdots \int f(u, v) \varphi(u) \, dx.$$

Proof. By definition,

$$\langle g, \varphi \rangle = \int_U g(u) \varphi(u) \, du.$$

By definition of g this becomes

$$\int_U \int_V f(u, v)\, dv\, \varphi(u)\, du,$$

and the desired result follows from the standard Fubini Theorem which allows interchange of the order of integration for L^1 functions.

Consistency Theorem 7.3c. Let f be L^1 on $\mathbb{R}^q$ and let g be the partial integral $\int_V f(u, v)\, dv$, as in the preceding theorem. Then the Fourier transform $\hat{g}(\hat{u})$ of g is simply the localization of $\hat{f}(t)$ to the subspace U. That is, if we use the notation $x = u + v$, $t = \hat{u} + \hat{v}$, as explained in Section 1, then

$$\hat{g}(\hat{u}) = \hat{f}(\hat{u}, 0).$$

Proof. We simply compute. By definition,

$$\hat{g}(\hat{u}) = \int_U e^{-2\pi i \hat{u} \cdot u} g(u)\, du$$

$$= \int_U \int_V e^{-2\pi i \hat{u} \cdot u} f(u, v)\, dv\, du.$$

Now, for any vector $x = u + v$ in $\mathbb{R}^q$, the inner product $(\hat{u}, 0) \cdot (u, v)$ is just $\hat{u} \cdot u$. Hence, in the term $e^{-2\pi i \hat{u} \cdot u}$, we can replace the '$u$' by '$x$' and (again using $du\, dv = dx$) obtain

$$\int_{\mathbb{R}^q} e^{-2\pi i (\hat{u}, 0) \cdot x} f(x)\, dx = \hat{f}(\hat{u}, 0). \qquad \text{Q.E.D.}$$

Note. Sometimes it is convenient to drop the 'hat' over the 'u', and write $\hat{g}(u)$ instead of $\hat{g}(\hat{u})$. Thus in the above proof we could just as well have written

$$\hat{g}(u) = \int_U e^{-2\pi i u \cdot \hat{u}} g(\hat{u})\, d\hat{u}.$$

Now, from the three Consistency Theorems above, we can easily infer what the correct definitions should be. From Theorem 7.3a we infer that a distribution T should be called 'integrable over $\mathbb{R}^q$', IF $\hat{T}(t)$ is continuous near $t = 0$, and then the value of $\int_{\mathbb{R}^q} T(x)\, dx$ should be defined to be $\hat{T}(0)$.

(Actually, the notion of a distribution being 'continuous near $t = 0$' also requires a definition, which will be given below.)

From Theorem 7.3b we infer that a distribution T should be called 'partially integrable over V' IF, for every test function $\varphi(u)$ on U, the product $T(u, v)\varphi(u)$ is integrable over $\mathbb{R}^q$. Again write $S(u) = \int_V T(u, v)\, dv$. Then, furthermore, the integral $\int_{\mathbb{R}^q} T(u, v)\varphi(u)\, dx$ *defines the action* $\langle S, \varphi \rangle$ of $S(u)$ on the test function $\varphi(u)$.

Finally, from Theorem 7.3c we infer that the Fourier transform $\hat{T}(u, v)$ should be called 'localizable to the subspace U' IF the original distribution

$T(u, v)$ is partially integrable dv. Then, if $S(u)$ denotes $\int_V T(u, v)\,dv$, the localization of $\hat{T}(u, v)$ to U should be $\hat{S}(u)$. Of course, by the Fourier Inversion Theorem, we can turn this around. Since $T^{\hat{}\check{}} = T^{\check{}\hat{}} = T$, we can replace $\hat{T}$ by T, and T by $\check{T}$. For the sake of harmony – in order to have all of our definitions apply originally to a distribution 'T' – this is what we shall do.

Now we simply repeat, under the more formal rubric of 'Definitions', what has just been said.

Definition. A tempered distribution T on $\mathbb{R}^q$ is called *continuous at* 0 if there is a neighborhood $\mathcal{N}$ of 0 and a continuous function f on $\mathcal{N}$ such that for all test functions φ with support $(\varphi) \subseteq \mathcal{N}$,

$$\langle T, \varphi \rangle = \langle f, \varphi \rangle = \int_{\mathbb{R}^q} f(x)\varphi(x)\,dx.$$

Then we define $T(0)$ to be $f(0)$.

(Since f is required to be continuous, it is trivial to verify that the value $f(0)$ is unique – i.e. that $T(0)$ is well defined.)

Now we give the definitions corresponding to the Consistency Theorems 7.3a, 3b and 3c. As noted above, these Consistency Theorems provide the motivation for the definitions which follow.

Definition (a). A tempered distribution T is called *integrable* (over $\mathbb{R}^q$) if its Fourier transform $\hat{T}$ is continuous at 0. Then we define $\int_{\mathbb{R}^q} T(x)\,dx$ by

$$\int \cdots \int_{\mathbb{R}^q} T(x)\,dx = \hat{T}(0).$$

In the next two definitions, we assume that $\mathbb{R}^q$ is the sum of two orthogonal complementary subspaces, $\mathbb{R}^q = U + V$, as above. We use the variables $\hat{u}, \hat{v}$ as above. Finally we assume that $T(x) = T(u, v)$ is a tempered distribution on $\mathbb{R}^q$, and that $S(u)$ is a tempered distribution on U.

Definition (b). We say that $T(u, v)$ is *partially integrable over* V and that $S(u) = \int_V T(u, v)\,dv$, if, for every open support test function $\varphi(u)$ on U, the product $T(u, v)\varphi(u)$ is integrable over $\mathbb{R}^q$ and

$$\langle S, \varphi \rangle = \int \cdots \int_{\mathbb{R}^q} T(u, v)\varphi(u)\,dv\,du.$$

(Again we recall that, by our conventions, dv du = dx.)

Definition (c). We say that $T(u, v)$ is *localizable to* U and write $S(u) = T(u, 0)$

if the inverse Fourier transform $\check{T}(u, v)$ is partially integrable over V and

$$\check{S}(u) = \int_V \check{T}(u, v)\, dv.$$

Notes. We could just as well use the direct Fourier transform instead of the inverse Fourier transform. Then the above displayed formula becomes

$$\hat{S}(u) = \int_V \hat{T}(u, v)\, dv.$$

For $\hat{S}(u)$ is just $\check{S}(-u)$, and similarly $\hat{T}(u, v) = \check{T}(-u, -v)$. Furthermore, as often noted, all integrals in distribution theory are taken 'in the positive sense', so that $\int_V \check{T}(-u, -v)\, dv = \int_V \check{T}(-u, v)\, dv$.

Because of its importance, we reemphasize that localization and partial integration are dual under the Fourier transform. That is, $T(u, v)$ is localizable to U if and only if $\check{T}(u, v)$ (or $\hat{T}(u, v)$) is partially integrable over V. When that holds,

$$[T(u, 0)]^{\vee} = \int_V \check{T}(u, v)\, dv,$$

and similarly if we replace all of the '$\vee$' by '$\wedge$'.

It is important to recall that the integral, partial integral, or localization of a tempered distribution T does not always exist. For some T it exists, and for other T it does not. Similarly, given a fixed distribution T, the partial integral may exist for some subspaces U, V and not exist for others. Likewise for the localization. The following example illustrates these points.

Example. Let $q = 2$, and use the conventional variables (x, y) to describe points in $\mathbb{R}^2$. Let U be the x-axis, and let V be the y-axis, with $u = \hat{u} = x$ and $v = \hat{v} = y$. Let

$$T(x, y) = 1 \cdot e^{-\pi y^2},$$

so that

$$\hat{T}(x, y) = \delta(x) \cdot e^{-\pi y^2}.$$

Then, clearly, $\hat{T}(x, y)$ is not continuous at $(0, 0)$ – i.e. the Dirac Delta Function is not continuous – so that $T(x, y)$ is not integrable over $\mathbb{R}^2$. However, $T(x, y)$ is partially integrable dy, and (as we would expect, since $\int_{-\infty}^{\infty} e^{-\pi y^2}\, dy = 1$) the partial integral is the constant function 1, *defined on the x-axis*. Let us work this out in terms of Definition (b).

We want to show that, for any test function $\varphi(x)$ defined on the x-axis,

$$\langle 1, \varphi \rangle = \iint_{\mathbb{R}^2} e^{-\pi y^2} \varphi(x)\, dy\, dx.$$

Now here the 'integral' over $\mathbb{R}^2$ is the distribution-theoretic integral given by Definition (a). But since $e^{-\pi y^2}\varphi(x)$ is integrable (in the traditional L^1 sense), we have by Consistency Theorem 7.3a that the distribution-theoretic integral and the ordinary integral coincide. Thus

$$\iint_{\mathbb{R}^2} e^{-\pi y^2}\varphi(x)\,dy\,dx = \int_{-\infty}^{\infty}\int_{-\infty}^{\infty} e^{-\pi y^2}\varphi(x)\,dy\,dx,$$

and since $\int_{-\infty}^{\infty} e^{-\pi y^2}dy = 1$, the last integral is

$$\int_{-\infty}^{\infty} \varphi(x)\,dx = \langle 1, \varphi \rangle,$$

as desired.

We leave it to the reader to verify that, on the other hand, $T(x, y) = 1 \cdot e^{-\pi y^2}$ is not partially integrable dx.

Now, passing to the Fourier transform $\hat{T}(x, y) = \delta(x) \cdot e^{-\pi y^2}$, it follows immediately that $\hat{T}(x, y)$ is localizable to the x-axis, and that we get $\hat{T}(x, 0) = \delta(x)$, just as we would expect. On the other hand, $\hat{T}(x, y)$ is not localizable to the y-axis.

Example. In the previous example, the 'total integral over $\mathbb{R}^q$' $(q = 2)$, once we got to it, was an ordinary classical integral. In other words, for that example, Definition (1) was superfluous. This is not always the case, as we now show.

Let $q = 1$, and let $T(x)$ be the distribution on $\mathbb{R}^1$ given by

$$T(x) = e^{2\pi i a x}, \quad a \neq 0.$$

Then $\hat{T}(t) = \delta(t - a)$, which is continuous and vanishes identically near $t = 0$ (although, of course, it is not continuous near $t = a$). Hence, by Definition (a),

$$\int_{-\infty}^{\infty} e^{2\pi i a x}\,dx = 0, \quad a \neq 0.$$

In this case the integral, although well defined according to Definition (a), has no classical meaning.

For applications, it is useful to have a direct description of the localization, i.e. a description not depending on the Fourier transform. We give two such descriptions. The first is necessary and sufficient – i.e. equivalent. The second is merely sufficient, but is useful in applications. First we need:

Definition of T_φ. For any open support test function $\varphi(u)$ on U, we let $T_\varphi(u, v)$ be the corresponding distribution on $\mathbb{R}^q$ given by

$$\langle T_\varphi(u, v), \psi(u, v) \rangle = \left\langle T(u, v), \int_U \psi(u - a, v)\varphi(a)\,da \right\rangle$$

for all $\psi(u, v) \in \mathscr{S}(\mathbb{R}^q)$.

Notes. We observe that the integral $\int_U \psi(u - a, v)\varphi(a)\, da$ above is an ordinary integral, in the sense of advanced calculus. It is easy to verify that T_φ, as so defined is a tempered distribution. Linearity is obvious. As for the second defining property of distributions, namely that $\psi_n \to \psi$ in $\mathscr{S}(\mathbb{R}^q)$ implies $\langle T, \psi_n \rangle \to \langle T, \psi \rangle$: if $\psi_n \to \psi$ in $\mathscr{S}(\mathbb{R}^q)$, then trivial estimates show that $\int_U \psi_n(u - a, v)\varphi(a)\, da \to \int_U \psi(u - a, v)\varphi(a)\, da$ in $\mathscr{S}(\mathbb{R}^q)$; thus the second property for T (since T is a distribution) implies the same for T_φ.

We emphasize that T_φ exists for any tempered distribution $T(u, v)$ and any test function $\varphi(u)$, whether T is localizable or not. However, the intuitive meaning of T_φ may be fairly well hidden. We will give a brief discussion of its intuitive content at the end of this section. For now, it seems more appropriate to press on and prove the two key theorems of this section.

Theorem 7.4 (First criterion for localization). Let $\mathbb{R}^q = U + V$ as above, and let $T(u, v)$ and $S(u)$ be tempered distributions on $\mathbb{R}^q$ and U, respectively. Let T_φ be as above. Then $T(u, v)$ is localizable to $S(u)$ on U if and only if, for all $\varphi \in \mathscr{S}(U)$, $T_\varphi(u, v)$ is continuous at $(0, 0)$ and

$$T_\varphi(0, 0) = \langle S, \varphi \rangle.$$

Proof of Theorem 7.4. First we must combine Definitions (c), (b) and (a). By Definition (c), $T(u, v)$ is localizable on U to $S(u)$ if and only if $\check{T}(u, v)$ is partially integrable over V and

$$\check{S}(u) = \int_V \check{T}(u, v)\, dv.$$

By definition (b), this holds if and only if, for every open support test function $\phi(u) \in \mathscr{S}(U)$, $\check{T}(u, v)\phi(u)$ is integrable over $\mathbb{R}^q$ and

$$\langle \check{S}(u), \phi(u) \rangle = \int_{\mathbb{R}^q} \check{T}(u, v)\phi(u)\, du\, dv.$$

(Here, of course, we have used the Fourier Inversion Theorem for $\mathscr{S}(U)$: since $\psi^{\wedge\vee} = \psi$, every test function ψ has the form ϕ for some φ, namely for $\varphi = \check{\psi}$.)

By Definition (a), the above reduces to

$$[\check{T}(u, v)\phi(u)]^{\wedge} \text{ is continuous at } (0, 0),$$

and

$$[\check{T}(u, v)\phi(u)]^{\wedge}(0, 0) = \langle \check{S}, \check{\phi} \rangle.$$

Now, since $\langle \check{S}, \phi \rangle = \langle S, \varphi^{\wedge\vee} \rangle = \langle S, \varphi \rangle$, this in turn reduces to

$$\langle S(u), \varphi(u) \rangle = [\check{T}(u, v)\phi(u)]^{\wedge}(0, 0),$$

together with the implicit statement that $[\check{T}(u, v)\phi(u)]^{\wedge}$ is continuous at $(0, 0)$.

This, then, is what the localizability of $T(u, v)$ to $S(u)$ on U actually means. Thus, to prove Theorem 7.4, it suffices to show that $[\check{T}(u, v)\phi(u)]^{\wedge} = T_\varphi$. Take any test function $\psi(u, v)$. Then

$$\langle [\check{T}(u, v)\phi(u)]^{\wedge}, \psi(u, v) \rangle = \langle \check{T}(u, v)\phi(u), \hat{\psi}(u, v) \rangle$$
$$= \langle \check{T}(u, v), \phi(u)\hat{\psi}(u, v) \rangle$$
$$= \langle T(u, v), [\phi(u)\hat{\psi}(u, v)]^{\vee} \rangle.$$

Now we do the ' $^{\vee}$ ' operation as the composition of two partial inverse Fourier transforms $F_u F_v$. From F_v, since $\phi(u)$ can be regarded as a constant, we get $\hat{\psi}^{\vee} = \psi$ in terms of its dependence on v. For F_u, we use the identity $(f * g)^{\wedge} = \hat{f}\hat{g}$, so that by the Fourier Inversion Theorem $(\hat{f}\hat{g})^{\vee} = f * g$. Hence the last expression above is equal to a convolution in terms of the u variable, namely

$$\left\langle T(u, v), \int_U \psi(u - a, v)\varphi(a)\, \mathrm{d}a \right\rangle. \qquad \text{Q.E.D.}$$

Now we relate the above criterion to the notion of 'cross section', as treated in Section 2. It turns out that 'cross section' is a special case of 'localizaton'. Thus the following provides a sufficient, but not necessary, condition for the localization to exist.

Note. We defined the cross sections $T_v(u)$ for any hyperspace $\{v = v_0 = \text{constant}\}$ parallel to the subspace U which is $\{v = 0\}$. The localization has been defined only for U. But this is just a detail. For the localization of $T(u, v)$ to $\{v = v_0\}$ would, of course, be defined as the localization of $T(u, v + v_0)$ to $\{v = 0\}$.

Theorem 7.5 (Second criterion for localization). Let $T(u, v)$ be a tempered distribution on $\mathbb{R}^q = U + V$ which has a family of cross sections $\{T_v(u)\}$ as defined in Section 2. Then, for each fixed $v_0 \in V$, the distribution $T(u, v + v_0)$ is localizable to U, and the localization is $T_{v_0}(u)$.

Proof. Since the definition of 'cross section' in Section 2 is homogeneous in v, we can, without loss of generality, take $v_0 = 0$.

We shall use the first criterion above to derive this one. Since the condition '$T_\varphi(u, v)$ is continuous at $(0, 0)$' involves only test functions supported in a neighborhood of $(0, 0)$, we can assume, in the arguments that follow, that $\psi(u, v)$ has compact support. We first observe that by property 2 in the Definition in Section 2, $\langle T_v(a), \varphi(a - u) \rangle_a$ is a continuous function of u and v. We will show that this function coincides with $\cdot T_\varphi(u, v)$, at least when applied to the test functions ψ with compact support.

Since this is one of the most delicate arguments we have encountered so far, we will include a note of explanation before each step. Now,

$$\langle\langle T_v(a), \varphi(a - u)\rangle_a, \psi(u, v)\rangle$$

is, since $\langle T_v(a), \varphi(a-u) \rangle_a$ is continuous and ψ has compact support, equal to

$$\int_{\mathbf{R}^q} \langle T_v(a), \varphi(a-u) \rangle_a \psi(u, v) \, du \, dv,$$

which, since $\psi(u, v)$ is constant in the variable a, and by the ordinary Fubini Theorem for interchanging multiple integrals, is equal to

$$\int_U \int_V \langle T_v(a), \varphi(a-u)\psi(u, v) \rangle_a \, dv \, du,$$

which by property 1 in the Definition in Section 2 is

$$\int_U \langle T_v(a), \varphi(a-u)\psi(u, v) \rangle_{a,v} \, du,$$

which by definition of the distribution '1' is

$$\langle 1(u), \langle T(a, v), \varphi(a-u)\psi(u, v) \rangle_{a,v} \rangle_u,$$

which by commutativity of the tensor product is

$$\langle T(a, v), \langle 1(u), \varphi(a-u)\psi(u, v) \rangle_u \rangle_{a,v},$$

which by definition of '1' is

$$\left\langle T(a, v), \int_U \varphi(a-u)\psi(u, v) \, du \right\rangle_{a,v},$$

which by a change of notation is

$$\left\langle T(u, v), \int_U \varphi(u-a)\psi(a, v) \, da \right\rangle_{u,v},$$

which upon setting $b = u - a$ becomes

$$\left\langle T(u, v), \int_U \varphi(b)\psi(u-b, v) \, db \right\rangle_{u,v},$$

which by definition of T_φ is

$$\langle T_\varphi(u, v), \psi(u, v) \rangle. \qquad\qquad \text{Q.E.D.}$$

Some concluding remarks. As already noted, the intuitive meaning of the distribution T_φ used in Theorem 7.4 may be fairly well hidden. It becomes clearer if we make some formal transformations (which here, since we are doing heuristics, we are not attempting to be rigorous). First we observe that, by the definition of translation for distributions, $\langle T(u, v), \psi(u-a, v) \rangle = \langle T(u+a, v), \psi(u, v) \rangle$. The weight factor $\varphi(a)$ is a constant in terms of u and v, and a little thought shows that we can take the integral outside and write the definition of T_φ, acting on a test function ψ, as

$$\langle T_\varphi(u, v), \psi(u, v) \rangle = \int_U \langle T(u+a, v)\varphi(a), \psi(u, v) \rangle \, da.$$

Now suppose, as we so often do, that we drop out the test function $\psi(u, v)$ and declare that two distribution-theoretic expressions are equal if they act in the same way on all test functions ψ. We are led to the formal expression

$$T_\varphi(u, v) = \int_U T(u + a, v)\varphi(a)\, da.$$

Thus T_φ is nothing more than a weighted average, with the weight factor $\varphi(a)$, of translates of the original distribution T.

Finally, our criterion for localizability of $T(u, v)$ to $S(u)$ is that, for all φ, $T_\varphi(u, v)$ be continuous at $(0, 0)$ and $T_\varphi(0, 0) = \langle S, \varphi \rangle$. Let's try this out on the last displayed formula above. Putting $u = v = 0$ in this formula, we get

$$T_\varphi(0, 0) = \int_U T(a, 0)\varphi(a)\, da,$$

which, combined with the formal symbolism,

$$\langle S, \varphi \rangle = \int_U S(a)\varphi(a)\, da,$$

suggests that $S(a) = T(a, 0)$. That is, S is the localization of $T(u, v)$ to the subspace $U = \{v = 0\}$, as desired.

Of course the above arguments are purely formal, whereas Theorem 7.4, as we have stated it, is rigorous and exact. However, the above arguments do perhaps convince us that Theorem 7.4 is what we should expect.

5 Multiplication and convolution

Let S and T be tempered distributions on $\mathbb{R}^q$. We are now in a position to give general definitions for the product ST and the convolution $S * T$.

Here we recall some points made in the introduction to this chapter. 'General' does not mean 'universal'. Sometimes the product and convolution do not exist! By a 'general' definition, we mean one which gets away from the extremely one-sided conditions set throughout Chapters 1–6. There, for the multiplication $g(x)T(x)$, we required that g be a C^∞ *function*. That is, we set extremely rigid restrictions on g, while allowing absolute freedom for T. Similarly for the convolution $S * T$ (in Chapter 2), we required that T have compact support. The definition given here is more balanced – treating both distributions S and T in a symmetrical manner. It is also more general, including the previous definitions as special cases. Finally, it includes cases which are important in analysis and its applications, but the previous definitions do not cover (cf. Section 6).

In setting down the general notions of multiplication and convolution, we will find that most of the work has already been done in Section 4. Consequently this section is very short. There is only one main theorem.

Two tempered distributions $S(x)$ and $T(x)$ are 'multiplicable' if and only if their Fourier transforms $\hat{S}(t)$ and $\hat{T}(t)$ are 'convolvable', and then we have the identity that we would expect:

$$[S(x)T(x)]\hat{} = \hat{S}(t) * \hat{T}(t).$$

The format for these developments is the following. We begin with two tempered distributions $S(x)$ and $T(x)$ on $\mathbb{R}^q$. Then we pass to the cartesian product $\mathbb{R}^q + \mathbb{R}^q = \mathbb{R}^{2q}$.

(We observe that here, even if our only interest were in one-dimensional distributions S and T, we should still have to work over the two-dimensional space $\mathbb{R}^2$.)

We begin with multiplication. Here a brief preview. We will take the tensor product $S(\sqrt{2}x)T(\sqrt{2}y)$, as defined in Section 3. (The reason for the '$\sqrt{2}$' will be made clear below.) Then, to define the product $(ST)(u)$, we take the localization of $S(\sqrt{2}x)T(\sqrt{2}y)$ to the subspace $U = \{y = x\}$, where U is parametrized by the vector $u = (y + x)/\sqrt{2}$. Now for the details.

It is convenient to fix the following notations. A point in $\mathbb{R}^{2q} = \mathbb{R}^q + \mathbb{R}^q$ will be written (x, y), where $x = (x_1, \ldots, x_q)$ denotes the first 'coordinate' in the cartesian product, and $y = (y_1, \ldots, y_q)$ denotes the second one. We stress that x and y are q-vectors: $x, y \in \mathbb{R}^q$.

Within $\mathbb{R}^{2q}$ we identify two subspaces of particular importance, namely

$$U = \{y = x\},$$
$$V = \{y = -x\}.$$

Then U and V are q-dimensional complementary orthogonal subspaces of $\mathbb{R}^{2q}$. We introduce canonical orthonormal coordinate systems corresponding to this decomposition by setting

$$u = \frac{y + x}{\sqrt{2}}, \qquad v = \frac{y - x}{\sqrt{2}},$$

so that

$$x = \frac{u - v}{\sqrt{2}}, \qquad y = \frac{u + v}{\sqrt{2}}.$$

Of course, the '$\sqrt{2}$' factors are necessary so that the u, v coordinate system will be orthonormal. Again, the coordinates u and v are q-vectors, corresponding to the orthogonal decomposition $\mathbb{R}^{2q} = U + V$.

Now we are ready to define the multiplication of distributions. Let S and T be arbitrary tempered distributions on $\mathbb{R}^q$. Then $S(\sqrt{2}x)$ and $T(\sqrt{2}x)$ are also tempered distributions. We form the tensor product $S(\sqrt{2}x)T(\sqrt{2}y)$.

(We remark that this tensor product always exists – cf. Section 3.)

Here we come to the operation which may or may not 'exist', and which determines whether the distributions S and T are multiplicable or not. We say that S and T are *multiplicable* if and only if the distribution $S(\sqrt{2}x)T(\sqrt{2}y)$

on $\mathbb{R}^{2q} = U + V$ is localizable to the subspace U – cf. Section 4. When this holds, we define the product $(ST)(u)$ to be this localization.

It is useful to work this out in terms of the variables u and v. Since $x = (u - v)/\sqrt{2}$ and $y = (u + v)/\sqrt{2}$, we have

$$S(\sqrt{2}x)T(\sqrt{2}y) = S(u - v)T(u + v),$$

(note that this is still a tensor product!) and the condition for multiplicability is that $S(u - v)T(u + v)$ be localizable to the subspace $U = \{v = 0\}$. The formula for the product becomes

$$(ST)(u) = S(u - v)T(u + v)\big|_{v = 0},$$

just as we would expect. Now we make this a formal definition.

Definition. Let S and T be arbitrary tempered distributions on $\mathbb{R}^q$. Let $S(u - v)T(u + v)$ be the tensor product on $\mathbb{R}^{2q} = U + V$ as described above. We say that S and T are *multiplicable* if $S(u - v)T(u + v)$ has a localization to the subspace U. When this holds, we define the product ST by

$$ST(u) = S(u - v)T(u + v)\big|_{v = 0}.$$

Note. It is easy to verify that this operation is commutative: that is, ST exists if and only if TS exists, and then $ST = TS$. For the tensor product is commutative (Theorem B in Section 3). Furthermore, the change of variable $v \leftrightarrow -v$ does not affect the localizability of the distribution $S(u - v)T(u + v)$ to the subspace $U = \{v = 0\}$ (cf. Theorem 7.4 in Section 4).

Now we turn to convolution. Again the definition requires a bit of preface. Begin with the classical case where f and g are integrable functions, and the convolution is given by an ordinary integral

$$(f * g)(u) = \int_{\mathbb{R}^q} f(u - w)g(w)\, dw.$$

This does not quite fit our format, since the variables $u - w$ and w do not correspond to orthonormal (or even to orthogonal) coordinate systems. So we make a change of variables. Let

$$v = 2w - u, \qquad w = \frac{u + v}{2}.$$

Then the q-dimensional volume element $dw = 2^{-q}\, dv$. The above integral becomes

$$(f * g)(u) = 2^{-q} \int_{\mathbb{R}^q} f\left(\frac{u - v}{2}\right) g\left(\frac{u + v}{2}\right) dv.$$

We almost have it! The vector variables $(u - v)/2$ and $(u + v)/2$ correspond to orthogonal (but not yet orthonormal) coordinates. As we have seen above, the correct orthonormal coordinates go with the variables $x = (u - v)/\sqrt{2}$

and $y = (u + v)/\sqrt{2}$. So, to get the proper expression in the last displayed integral above, we simply take the product $f(x/\sqrt{2})g(y/\sqrt{2})$.

Thus, consider two arbitrary tempered distributions S and T on $\mathbb{R}^q$. Then $S(x/\sqrt{2})$ and $T(x/\sqrt{2})$ are also tempered distributions. We form the tensor product $S(x/\sqrt{2})T(y/\sqrt{2})$. As we have seen,

$$S(x/\sqrt{2})T(y/\sqrt{2}) = S\left(\frac{u-v}{2}\right)T\left(\frac{u+v}{2}\right).$$

This leads us to the following.

Definition. Let S and T be arbitrary tempered distributions on $\mathbb{R}^q$. Let $S((u - v)/2)T((u + v)/2)$ be the tensor product on $\mathbb{R}^{2q} = U + V$ as described above. We say that S and T are *convolvable* if $S((u - v)/2)T((u + v)/2)$ is partially integrable over the subspace V. When this holds, we define the convolution $S * T$ by

$$(S * T)(u) = 2^{-q}\int_V S\left(\frac{u-v}{2}\right)T\left(\frac{u+v}{2}\right)dv.$$

The next theorem asserts that there is a complete symmetry between multiplication and convolution under the Fourier transform. This fact allows us to turn any problem concerning convolution into a corresponding problem for multiplication, and vice versa.

Theorem 7.6. Let S and T be arbitrary tempered distributions on $\mathbb{R}^q$, and let $\hat{S}$ and $\hat{T}$ be their Fourier transforms. Then S and T are multiplicable if and only if $\hat{S}$ and $\hat{T}$ are convolvable. When this holds, then

$$(ST)\hat{} = \hat{S} * \hat{T}.$$

Note. Of course, the same result follows immediately for the inverse Fourier transform, since $\check{S}(t) = \hat{S}(-t)$. Then, by the Fourier Inversion Theorem, we can if we wish turn the whole thing around: S and T are convolvable if and only if $\hat{S}$ and $\hat{T}$ are multiplicable, and then $(S * T)\hat{} = \hat{S}\hat{T}$.

Proof of Theorem 7.6. It is convenient to use the variables x and y rather than u and v – cf. the discussions preceding the definitions of multiplication and convolution. We recall that $u = (y + x)/\sqrt{2}$, $v = (y - x)/\sqrt{2}$, and $x = (u - v)/\sqrt{2}$, $y = (u + v)/\sqrt{2}$.

Begin with the tensor product $S(u - v)T(u + v)$ which appears in the definition of multiplication. As we have seen, this is just $S(\sqrt{2}x)T(\sqrt{2}y)$.

Now, since the variable x is q-dimensional,

$$[S(\sqrt{2}x)]\hat{} = (\sqrt{2})^{-q} \cdot \hat{S}(x/\sqrt{2}).$$

Similarly,

$$[T(\sqrt{2}y)]\hat{} = (\sqrt{2})^{-q} \cdot \hat{T}(y/\sqrt{2}).$$

We recall that, by Theorem 7.1 in Section 3, the Fourier transform of a tensor product is the corresponding tensor product of the Fourier transforms. Thus,

$$[S(\sqrt{2}x)][T(\sqrt{2}y)]\hat{} = 2^{-q}\hat{S}(x/\sqrt{2})\hat{T}(y/\sqrt{2}).$$

But, as we have seen, $\hat{S}(x/\sqrt{2})\hat{T}(y/\sqrt{2})$ is $\hat{S}((u-v)/2)\hat{T}((u+v)/2)$. This is precisely the product which appears in the definition of the convolution $\hat{S} * \hat{T}$. To summarize what we have so far:

$$[S(u-v)T(u+v)]\hat{} = 2^{-q}\hat{S}\left(\frac{u-v}{2}\right)\hat{T}\left(\frac{u+v}{2}\right).$$

We are nearly finished. By definition, the product $(ST)(u)$ is the localization of $S(u-v)T(u+v)$ to the subspace U. By definition, the convolution $(\hat{S} * \hat{T})(u)$ is the partial integral of $2^{-q}\hat{S}((u-v)/2)\hat{T}((u+v)/2)$ over the subspace V. Finally, by the previous Definitions (b) and (c) from Section 4, localization and partial integration are symmetrical under the Fourier transform. (See the Notes following Definition (c) in Section 4.) Since $2^{-q}\hat{S}((u-v)/2)\hat{T}((u+v)/2)$ is the Fourier transform of $S(u-v)T(u+v)$,

$$[(ST)(u)]\hat{} = (\hat{S} * \hat{T})(u),$$

as desired.

6 Examples

Again, it is useful to pause briefly and recall the ground we have covered. At the beginning of Section 5, we promised to set down 'general' definitions of multiplication and convolution – i.e. definitions which, while not universal, would nevertheless cover the cases which occur routinely in analysis and its applications. We then set down those definitions and proved that they are symmetrical under the Fourier transform (Theorem 7.6). Now we must attempt to show that the promise has been kept – i.e. that these definitions do encompass the cases of multiplication and convolution which occur most routinely in analysis.

We will begin with multiplication. This depends on tensor products, as developed in Section 3, and localization, developed in Section 4. However, in all of the important cases, we shall find that the much easier concept of 'distributions depending on a parameter' (Section 2) actually suffices. The complications in Section 4 are necessary to achieve symmetry under the Fourier transform. Mercifully, we will see that the complications are mostly over, and that everything goes fairly smoothly from now on.

In Theorem 7.5 of Section 4, we showed that any distribution $T(u, v)$ which satisfies the conditions of Section 2 is localizable. For convenience, we recall the conditions set down in Section 2. These conditions are that $T(u, v)$ has a family of 'cross sections' $T_v(u)$, which are tempered distributions on U, such that:

Property 1. For each open support test function $\psi(u, v)$,

$$\langle T(u, v), \psi(u, v)\rangle = \int_V \langle T_v(u), \psi(u, v)\rangle \, dv.$$

Property 2. If $v \to v_0$ in V and $\varphi_n \to \varphi$ in $\mathscr{S}$, then

$$\langle T_v(u), \varphi_n(u)\rangle \to \langle T_{v_0}(u), \varphi(u)\rangle.$$

These, then, are the conditions we will have to verify in all of the 'multiplication' examples below. Following the definition of multiplication from Section 5, the conditions will be applied to distributions of the form $T(u, v) = f(u - v)g(u + v)$. Once these conditions are verified, the localizability of $f(u - v)g(u + v)$ to $U = \{v = 0\}$ follows. By definition, this means that f and g are multiplicable.

Example 1. (Multiplication of an L^p function by an L^r function.) Let us begin with the ordinary pointwise multiplication of a function $f(x)$ in L^p by a function $g(x)$ in L^r.

Now the product $f(x)g(x)$ may not be a distribution! A typical counter-example involves the square of an L^1 function. Thus, let $f(x)$ on $\mathbb{R}^1$ be $x^{-2/3}$ for $0 < x \leqslant 1$, and 0 elsewhere. Then $f(x) \in L^1$, but $[f(x)]^2$ has an infinite positive mass (with no negative mass to balance it) supported on a compact interval. Such infinite positive masses do not correspond to distributions.

When should $f(x)g(x)$ be a distribution? The classical case is when $(1/p) + (1/r) \leqslant 1$. For by a well known extension of Hölder's inequality ([DUS], chap. 6, Exercises):

$$\text{if } 1/p + 1/r = 1/t, t \geqslant 1, \text{ then } \|fg\|_t \leqslant \|f\|_p \|g\|_r. \tag{H}$$

In the special case when $(1/p) + (1/r) = 1$, we have Hölder's inequality:

$$\text{if } 1/p + 1/r = 1, \text{ then } \|fg\|_1 \leqslant \|f\|_p \|g\|_r.$$

Now we show that, as expected, when $(1/p) + (1/r) \leqslant 1$, then f and g are multiplicable in the sense of distribution theory.

Theorem 7.7. Assume that $1 \leqslant p \leqslant \infty$, $1 \leqslant r < \infty$ (strict inequality), and $(1/p) + (1/r) \leqslant 1$. Let $f \in L^p$ and $g \in L^r$. Then the corresponding distributions T_f and T_g are multiplicable.

Proof. As in the inequality (H) above, we write $(1/p) + (1/r) = (1/t)$, $t \geqslant 1$.

Let $T(u, v) = f(u - v)g(u + v)$. Then, since the test functions $\varphi(u, v)$ are rapidly decreasing, the integral of $\int_{\mathbb{R}^{2q}} T(u, v)\varphi(u, v) \, du \, dv$ exists in the classical sense as a Lebesque integral and gives rise to a tempered distribution. We must now verify that $T(u, v)$ has a family of cross sections $T_v(u)$, as described above. First we define the cross section. The definition is obvious.

For each fixed v, $T_v(u) = f(u - v)g(u + v)$. Then $T_v(u) \in L^t$ by (H) above, and hence is a tempered distribution.

Now we verify property 1 of the cross section:

For any open support test function $\psi(u, v)$, $f(u - v)g(u + v)\psi(u, v) \in L^1(U + V)$. Hence we can apply Fubini's Theorem to get the following:

$$\int_V \langle T_v(u), \psi(u, v) \rangle \, dv = \int_V \int_U f(u - v)g(u + v)\psi(u, v) \, du \, dv$$

$$= \int_{U+V} f(u - v)g(u + v)\psi(u, v) \, du \, dv$$

$$= \langle T(u, v), \psi(u, v) \rangle.$$

Now we verify property 2.

First we avoid the extreme case and assume that $p < \infty$. Now translation in L^p and L^r are continuous in terms of the L^p/L^r norms, respectively. And the inequality $\|fg\|_t \leqslant \|f\|_p \|g\|_r$ then implies that the product varies continuously in $L^t(U)$ as v varies. This is more than enough. We need to know that $\langle T_v(u), \varphi_n(u) \rangle \to \langle T_{v_0}(u), \varphi(u) \rangle$ as $v_0 \to v$ in V and $\varphi_n \to \varphi$ in $\mathscr{S}(U)$. By the Hölder inequality (cf. above), it would suffice if φ_n approached φ in the $L^{t'}$ norm $((1/t) + (1/t') = 1)$. But the fact that $\varphi_n \to \varphi$ in $\mathscr{S}(U)$ (with its very severe topology), trivially implies that $\varphi_n \to \varphi$ in $L^{t'}$.

For the case where $p = \infty$ (but $r < \infty$), we have $(1/p) + (1/r) = (1/r)$, so that $t = r$. Now the translation $f(u) \to f(u - v)$ is not continuous in the norm of $L^\infty(U)$. However, since $r < \infty$, translation *is* continuous in $L^r(U)$. This suffices. To show the argument, let τ_a be the translation operation $\tau_a[f(u)] = f(u - a)$. Then

$$f(u - v)g(u + v) = \tau_v[f(u)(\tau_{-2v}[g(u)])].$$

Now the translation τ_{-2v} is continuous in $L^r(U)$, multiplication by the bounded function $f(u)$ is continuous in terms of $L^r(U)$, and the translation τ_v is continuous in $L^r(U)$. Hence, as v varies, $f(u - v)g(u + v)$ varies continuously in $L^r(U)$.

Once the continuity of $f(u - v)g(u + v)$ in $L^r(U)$ has been established, the rest of the proof goes as before.

Example 2. (Multiplication: the standard case.) As expected, the new definition of multiplication includes the special case treated earlier, in Chapters 5 and 6.

Proposition. Let $g(x)$ be a C^∞ function which is slowly increasing together with its derivatives. Let $S(x)$ be an arbitrary tempered distribution. Then g

and S are multiplicable. Furthermore, the product of gS (by the new definition) coincides with that defined in Chapters 5 and 6.

Proof. We must define a family of cross sections and verify the properties 1 and 2. For each fixed v, the cross section $T_v(u)$ is defined in the obvious way. Namely: $\langle T_v(u), \varphi(u) \rangle = \langle g(u-v)S(u+v), \varphi(u) \rangle$, where the product $g(u-v)S(u+v)$ ($v = $ constant) is in the sense of Chapters 5 and 6.

We verify property 1:

$$\int_V \langle T_v(u), \psi(u,v) \rangle \, dv = \int_V \langle g(u-v)S(u+v), \psi(u,v) \rangle_u \, dv$$
$$= \langle 1(v), \langle g(u-v)S(u+v), \psi(u,v) \rangle_u \rangle_v,$$

which by definition of the tensor product is

$$\langle 1(v)g(u-v)S(u+v), \psi(u,v) \rangle,$$

which by the Consistency Theorem for multiplication reduces to

$$\langle g(u-v)S(u+v), \psi(u,v) \rangle.$$

Now we verify property 2:

Again we need to show that $\langle T_v(u), \varphi_n(u) \rangle \to \langle T_{v_0}(u), \varphi(u) \rangle$ as $v \to v_0$ in V and $\varphi_n \to \varphi$ in $\mathscr{S}(U)$.

Now $\langle T_v(u), \varphi_n(u) \rangle = \langle g(u-v)S(u+v), \varphi_n(u) \rangle = \langle S(u+v), g(u-v)\varphi_n(u) \rangle = \langle S(u), g(u-2v)\varphi_n(u-v) \rangle$. Thus we must study the behavior of

$$\langle S(u), g(u-2v)\varphi_n(u-v) \rangle$$

as $v \to v_0$ and $\varphi_n \to \varphi$.

It is trivial to verify that, since g is slowly increasing together with its derivatives, and since $\varphi_n \to \varphi$ in $\mathscr{S}(U)$, then

$$g(u-2v)\varphi_n(u-v) \to g(u-2v_0)\varphi(u-v_0)$$

in $\mathscr{S}(U)$. Thus, since S is a distribution, $\langle S(u), g(u-2v)\varphi_n(u-v) \rangle \to \langle S(u), g(u-2v_0)\varphi(u-v_0) \rangle$ as $v \to v_0$ and $\varphi_n \to \varphi$. This proves that g and S are multiplicable.

To show that gS (new definition) coincides with gS as defined in Chapters 5 and 6, we simply observe the following. The cross sections $T_v(u) = g(u-v)S(u+v)$ are already in the sense of Chapters 5 and 6. For $v = 0$, the cross section becomes $g(u)S(u)$. Finally, Theorem 7.5 guarantees that the localization to $\{v = 0\}$ coincides with the cross section for $v = 0$.   Q.E.D.

We give two more examples, whose proofs are left to the reader. Neither of these examples is covered by the definition of multiplication given in Chapters 5 and 6.

Example 3. (Multiplication of a continuous function and a finite measure.) Let f be a continuous slowly increasing function, and let μ be a finite complex measure. Then f and μ are multiplicable.

Example 4. (Multiplication involving functions which are C^n but not C^∞.) Let f be a C^n function which is slowly increasing together with all of its derivatives of order $\leqslant n$. Let μ be a finite complex measure. Then f is multiplicable with any mixed partial derivative $\mu^{(\alpha)}$ of μ of order $\leqslant n$.

Convolution. We will see that, perhaps surprisingly, convolution is easier to deal with than multiplication. For convolution, unlike multiplication, does not require the idea of 'cross sections'. We will be able to proceed directly from the idea of partial integration, defined in Section 4, and the definition of convolution given in Section 5.

Example 5. (Convolution of an L^p function with an L^r function.) In measure theory, the convolution of an L^p function with an L^r function exists whenever $(1/p) + (1/r) \geqslant 1$. Namely, as an application of the Riesz Convexity Theorem ([DUS], chap. 6, Exercises) we have:

$$\text{if } 1/p + 1/r = 1 + 1/t, \ t \geqslant 1, \text{ then } \|f*g\|_t \leqslant \|f\|_p \|g\|_r. \tag{R}$$

Important special cases are when $p = 1$, so that $t = r$, and when $(1/p) + (1/r) = 1$, so that $t = \infty$. Thus for $p = 1$: convolution with an L^1 function f is always possible, and

$$\|f*g\|_r \leqslant \|f\|_1 \|g\|_r.$$

When $(1/p) + (1/r) = 1$, we obtain

$$\|f*g\|_\infty \leqslant \|f\|_p \|g\|_r.$$

Since the convolution $f * g$ involves an integral, this last result is an extension of Hölder's inequality. It is also an easy consequence of Hölder's inequality.

Remark. As stated in the introduction to this chapter, the key weakness in our earlier treatment of convolution (Chapter 2) was that we required one of the two distributions to have compact support. Hence the whole measure-theoretic approach to convolution, as outlined above, was left out. This we now remedy.

Proposition. If $f \in L^p$, $g \in L^r$ and $1/p + 1/r \geqslant 1$, then the distributions T_f and T_g are convolvable.

Proof. We need to show that $f((u-v)/2)g((u+v)/2)$ is partially integrable over V. This means, by definition, that, for any test function $\varphi \in \mathscr{S}(U)$, $f(u-v)/2)g((u+v)/2)\varphi(u)$ is integrable over $\mathbb{R}^{2q}$.

We shall show that, in fact, $f((u-v)/2)g((u+v)/2)\varphi(u) \in L^1(U+V)$. By Consistency Theorem 7.3a, any function in L^1 is 'integrable' in the sense of distribution theory. Moreover, its 'ordinary L^1-integral' coincides with its distribution-theoretic integral.

To show that $f((u-v)/2)g((u+v)/2)\varphi(u) \in L^1(U+V)$. We need to show that

$$\int_U \int_V f\left(\frac{u-v}{2}\right) g\left(\frac{u+v}{2}\right) \varphi(u) \, dv \, du$$

exists as an ordinary Lebesque integral. We can write this as

$$2^q \int_U (f*g)(u)\varphi(u) \, du,$$

where

$$(f*g)(u) = 2^{-q} \int_V f\left(\frac{u-v}{2}\right) g\left(\frac{u+v}{2}\right) dv.$$

Thus $f*g$ is just the ordinary, Lebesque-theoretic convolution of f and g. By the inequality (R) above, $\|f*g\|_t \leqslant \|f\|_p \|g\|_r$, where $(1/p)+(1/r) = 1+(1/t)$. Therefore, by the Hölder inequality, to prove the L^1-integrability of $(f*g)(u)\varphi(u)$, it would suffice to have $\varphi \in L^{t'}(U)$, $(1/t)+(1/t') = 1$. But since φ is a test function – and hence rapidly decreasing – the condition $\varphi \in L^{t'}(U)$ is satisfied with room to spare. Q.E.D.

Example 6. (Convolution: the standard case.) We now show that the new definition of convolution includes the special case treated earlier in Chapter 2. We recall that, in this special case, we assumed that one of the two distributions had compact support. Thus we want to show that if S is an arbitrary tempered distribution and T is a distribution with compact support, then S and T are convolvable in the new sense defined in this chapter.

One approach would be to show that, if T has compact support, then the Fourier transform $\hat{T}$ is C^∞ and slowly increasing together with its derivatives. Then, by Example 2, $\hat{S}$ and $\hat{T}$ are multiplicable, and by Theorem 7.6 this implies that S and T are convolvable. However, we prefer a direct approach.

The following lemma is of some interest in its own right.

Main Lemma. If $T \in \mathscr{S}'$ and $\varphi \in \mathscr{S}$, then φT is integrable. Furthermore, the integral of φT equals $\langle T, \varphi \rangle$.

We begin with two minor lemmas.

Lemma. $(\varphi T)\hat{}(t) = \langle T(x), \varphi(x)e^{-2\pi i t x} \rangle$.

Proof of lemma. Let ψ be any test function. Then $\langle (\varphi T)\hat{}, \psi \rangle = \langle \varphi T, \hat{\psi} \rangle = \langle T, \varphi\hat{\psi} \rangle$. We need to prove that $\langle T, \varphi\hat{\psi} \rangle = \int \langle T(x), \varphi(x)e^{-2\pi i t x} \rangle \psi(t) \, dt$.

$$\int \langle T(x), \varphi(x)e^{-2\pi i t x} \rangle \psi(t) \, dt = \langle 1(t)T(x), \varphi(x)e^{-2\pi i t x}\psi(t) \rangle,$$

which by commutativity of the tensor product is

$$\langle T(x)1(t),\, \varphi(x)e^{-2\pi itx}\psi(t)\rangle = \langle T(x),\, \varphi(x)\hat{\psi}(x)\rangle. \qquad \text{Q.E.D.}$$

Lemma. $\langle T(x),\, \varphi(x)e^{-2\pi itx}\rangle$ is continuous in t. For $t = 0$, its value is $\langle T, \varphi\rangle$.

Proof of lemma. As t varies, $\varphi(x)e^{-2\pi itx}$ varies continuously in the topology of $\mathcal{S}$. Thus, since T is a distribution, $\langle T(x),\, \varphi(x)e^{-2\pi itx}\rangle$ varies continuously in t. Clearly, for $t = 0$, its value is $\langle T, \varphi\rangle$.

Proof of Main Lemma. The two (minor) lemmas together immediately imply the Main Lemma. For, by definition, the integral of φT is the value of its Fourier transform $(\varphi T)\hat{\,}$ at $t = 0$.

Proposition. If S and T are tempered distributions, and T has compact support, then S and T are convolvable. The convolution $S * T$ (new definition) coincides with that defined in Chapter 2.

Proof. We need to show that, for every test function $\varphi \in \mathcal{S}(U)$, $S((u - v)/2)T((u + v)/2)\varphi(u)$ is integrable. Since T has compact support, there exists a mesa function $\psi \in \mathcal{S}$ such that $\psi(x) \equiv 1$ over the support of T. Thus,

$$S\left(\frac{u-v}{2}\right)T\left(\frac{u+v}{2}\right)\varphi(u) = S\left(\frac{u-v}{2}\right)T\left(\frac{u+v}{2}\right)\varphi(u)\psi\left(\frac{u+v}{2}\right).$$

Now, since

$$\varphi(u)\psi\left(\frac{u+v}{2}\right) \in \mathcal{S}(U+V)$$

and

$$S\left(\frac{u-v}{2}\right)T\left(\frac{u+v}{2}\right) \in \mathcal{S}'(U+V),$$

we have by the Main Lemma above that

$$S((u-v)/2)T((u+v)/2)\varphi(u)\psi((u+v)/2)$$

is integrable. Thus S and T are convolvable.

To show that $S * T$ (new definition) coincides with $S * T$ as defined in Chapter 2, we argue as follows. The new convolution

$$\langle S * T, \varphi\rangle = 2^{-q}\int_{\mathbb{R}^{2q}} S\left(\frac{u-v}{2}\right)T\left(\frac{u+v}{2}\right)\varphi(u)\psi\left(\frac{u+v}{2}\right)\,dv\,du,$$

which by the Main Lemma is

$$2^{-q}\left\langle S\left(\frac{u-v}{2}\right)T\left(\frac{u+v}{2}\right),\, \varphi(u)\psi\left(\frac{u+v}{2}\right)\right\rangle,$$

which by the change of variable $x = (u-v)/2$, $y = -(u+v)/2$, $dx\,dy = 2^{-q}\,du\,dv$, is

$$\langle S(x)T(-y),\, \varphi(x-y)\psi(-y)\rangle,$$

which, since the mesa function $\psi(y) \equiv 1$ on the support of T, is

$$\langle S(x), \langle T(-y), \varphi(x - y) \rangle \rangle,$$

which by definition is

$$\langle S * T, \varphi \rangle$$

(as in Chapter 2).

7 The Fubini Theorem and related results

By a 'Fubini Theorem', we mean any result, at whatever level of generality, that allows the interchange of order in multiple integrals. Let $\mathbb{R}^q = U + V$ as above. The standard measure-theoretic version of Fubini's Theorem asserts that, if $f(x) = f(u, v)$ is integrable over $\mathbb{R}^q$, then

$$\int_{\mathbb{R}^q} f(x)\, dx = \int_U \int_V f(u, v)\, dv\, du.$$

Our objective here is to extend this result to the more general setting of distribution theory.

The classical Fubini Theorem (for ordinary functions) will be assumed. It is proved in virtually every book on measure theory. In fact, this classical result will be used several times in our proof. But, of course, for the more general distribution-theoretic results, we do give the proofs. They are not terribly hard, but not altogether trivial either.

Perhaps surprisingly, our first version of 'Fubini's Theorem' involves localization rather than partial integration.

Theorem 7.8 (Fubini's Theorem for localization). Write $\mathbb{R}^q = U + V$ as above. Let $T(u, v)$ be a tempered distribution which is continuous at $(0, 0)$, and suppose that $T(u, v)$ is localizable to $S(u)$ on U. Then $S(u)$ is continuous at $u = 0$, and $S(0) = T(0, 0)$.

Proof. To say that $T(u, v)$ is continuous at $(0, 0)$ means that there exists a continuous function $f(u, v)$ which coincides with $T(u, v)$ in a neighborhood of $(0, 0)$. To apply the assumption that $T(u, v)$ is localizable to $S(u)$, we use the first criterion for localization, Theorem 7.4, in Section 4. Thus, with the notations of Theorem 7.4: T_φ is the distribution on $\mathbb{R}^q = U + V$ defined by

$$\langle T_\varphi, \psi \rangle = \left\langle T(u, v), \int_U \psi(u - a, v)\varphi(a)\, da \right\rangle$$

for all ψ in $\mathscr{S}(\mathbb{R}^q)$. Then, for all φ in $\mathscr{S}(U)$, T_φ is continuous at $(0, 0)$, and $T_\varphi(0, 0) = \langle S, \varphi \rangle$.

Now to prove the theorem it suffices to show that $S(u) = f(u, 0)$ near $u = 0$. Thus, take a test function $\varphi(u)$ with small support around $u = 0$, and let us show that $\langle S(u), \varphi(u) \rangle = \langle f(u, 0), \varphi(u) \rangle$.

We know that

$$\langle S(u), \varphi(u) \rangle = T_\varphi(0, 0)$$

and

$$\langle f(u, 0), \varphi(u) \rangle = \int_U f(u, 0)\varphi(u) \, du.$$

Hence we need to show that, for all φ as above,

$$T_\varphi(0, 0) = \int_U f(u, 0)\varphi(u) \, du.$$

To see what T_φ is at $(u, v) = (0, 0)$, take a test function $\psi(u, v)$ with support near $(0, 0)$. Then, by definition of T_φ,

$$\langle T_\varphi, \psi \rangle = \left\langle T(u, v), \int_U \psi(u - a, v)\varphi(a) \, da \right\rangle.$$

Now, since $\psi(u, v)$ has a small support near $(0, 0)$ and $\varphi(u)$ has a small support near $u = 0$, the test function $\int_U \psi(u - a, v)\varphi(a) \, da$ has a small support near $(u, v) = (0, 0)$. Hence, since $T(u, v) = f(u, v)$ near $(0, 0)$, the above becomes

$$\int_{\mathbb{R}^q} f(u, v) \int_U \psi(u - a, v)\varphi(a) \, da \, du \, dv = \int f(u, v)\psi(u - a, v)\varphi(a) \, da \, du \, dv$$

$$= \int f(u, v)\psi(u - a, v)\varphi(a) \, du \, da \, dv$$

$$= \int f(u + a, v)\psi(u, v)\varphi(a) \, du \, da \, dv$$

$$= \int f(u + a, v)\psi(u, v)\varphi(a) \, da \, du \, dv$$

$$= \left\langle \int_U f(u + a, v)\varphi(a) \, da, \psi(u, v) \right\rangle.$$

(All of the above integrals involve continuous functions on a compact set, and hence the traditional Fubini Theorem – for ordinary integrals – applies to them.) Therefore

$$T_\varphi(u, v) = \int_U f(u + a, v)\varphi(a) \, da$$

near $(u, v) = (0, 0)$ and hence

$$T_\varphi(0, 0) = \int_U f(a, o)\varphi(a) \, da. \qquad \text{Q.E.D.}$$

As an immediate corollary, we deduce:

Theorem 7.8a (First Fubini Theorem for integration). Let $T(u, v)$ be integrable over $\mathbb{R}^q$ and also partially integrable over V to $S(u)$. Then $S(u)$ is integrable over U to the same value. In symbols:

$$S(u) = \int_V T(u, v)\, dv \quad \text{(by definition)},$$

and

$$\int_U S(u)\, du = \int_{\mathbb{R}^q = U + V} T(u, v)\, dv\, du.$$

Proof. By a slight abuse of language, this is simply the Fourier transform of the previous theorem. Thus $T(u, v)$ is integrable over $\mathbb{R}^q$ if and only if $\hat{T}(u, v)$ is continuous at $(0, 0)$, and $\int_{\mathbb{R}^q} T(u, v) = \hat{T}(0, 0)$. Similarly, $T(u, v)$ is partially integrable dv to $S(u)$ if and only if $\hat{T}(u, v)$ is localizable to $\hat{S}(u)$. Finally, $\int_U S(u)\, du = \hat{S}(0)$. Hence this result follows immediately from the previous one.

We would like to extend Theorem 7.8a to the case where $T(u, v)$ is merely partially integrable, instead of over $\mathbb{R}^q$. For this we need the following.

Theorem 7.9 (Variable Constants Theorem). Let $T(u, v)$ be partially integrable over V and $g(u)$ be a C^∞ function which is slowly increasing together with all of its derivatives. Then $g(u)T(u, v)$ is partially integrable over V and

$$\int_V g(u)T(u, v)\, dv = g(u) \int_V T(u, v)\, dv.$$

Proof. Take a test function $\varphi(u) \in \mathscr{S}(U)$ and consider $\int_{\mathbb{R}^q} \varphi(u)g(u)T(u, v)\, dv\, du$. Since both $\varphi(u)$ and $\varphi(u)g(u)$ are test functions in $\mathscr{S}(U)$, and since $T(u, v)$ is partially integrable over V, the integral exists and it can be viewed either as

$$\left\langle \int_V g(u)T(u, v)\, dv,\ \varphi(u) \right\rangle$$

or as

$$\left\langle \int_V T(u, v)\, dv,\ \varphi(u)g(u) \right\rangle,$$

which, since g is a C^∞ function, is

$$\left\langle g(u) \int_V T(u, v)\, dv,\ \varphi(u) \right\rangle.$$

Theorem 7.10 (Strong Fubini Theorem). Let $\mathbb{R}^q = U + V$ as above, and let the subspace V have the further decomposition $V = V_1 + V_2$. Suppose $T(u, v) = T(u, v_1, v_2)$ is partially integrable over V and also partially integrable over V_2. Then $\int_{V_2} T(u, v_1, v_2)\, dv_2$ is partially integrable over V_1 and

$$\int_{V_1} \int_{V_2} T(u, v_1, v_2)\, dv_2\, dv_1 = \int_V T(u, v)\, dv.$$

Proof. To show that $\int_{V_2} T(u, v_1, v_2)\, dv_2$, a distribution on $U + V_1$, is partially integrable over V_1, we need to show that, for any $\varphi \in \mathscr{S}(U)$,

$$\varphi(u) \cdot \int_{V_2} T(u, v_1, v_2)\, dv_2$$

is integrable over $U + V_1$. To show that

$$\int_{V_1} \int_{V_2} T(u, v_1, v_2)\, dv_2\, dv_1 = \int_V T(u, v)\, dv,$$

we need to show that

$$\int_{U+V_1} \varphi(u) \int_{V_2} T(u, v_1, v_2)\, dv_2\, dv_1\, du = \int_{\mathbb{R}^q} \varphi(u) T(u, v)\, dv\, du.$$

We now prove these statements.

Consider $\varphi(u) T(u, v_1, v_2)$. We will apply the First Fubini Theorem (Theorem 7.8a) to the complementary subspaces $U + V_1$ and V_2:

$$\int_{U+V_1} \int_{V_2} \varphi(u) T(u, v_1, v_2)\, dv_2\, dv_1\, du = \int_{\mathbb{R}^q} \varphi(u) T(u, v)\, dv\, du.$$

(Note that this differs from the above displayed formula, since here $\varphi(u)$ is inside the dv_2 integral.) To justify this last formula, we need to show that $\varphi(u) T(u, v)$ is integrable over $\mathbb{R}^q$ and also partially integrable over V_2. It is integrable on $\mathbb{R}^q$ because T is partially integrable over V. It is also partially integrable over V_2: for $\varphi(u)1(v_1)$ is a C^∞ function which is slowly increasing together with its derivatives, and T is partially integrable over V_2. Hence, by the Variable Constants Theorem, $\varphi(u)1(v_1)T(u, v_1, v_2) = \varphi(u)T(u, v_1, v_2)$ is partially integrable over V_2, and

$$\int_{V_2} \varphi(u) T(u, v_1, v_2)\, dv_2 = \varphi(u) \int_{V_2} T(u, v_1, v_2)\, dv_2.$$

Now to finish the proof. We begin with

$$\int_{U+V_1} \varphi(u) \int_{V_2} T(u, v_1, v_2)\, dv_2\, dv_1\, du,$$

which by the Variable Constants Theorem is

$$\int_{U+V_1} \int_{V_2} \varphi(u) T(u, v_1, v_2)\, dv_2\, dv_1\, du,$$

which by the First Fubini Theorem is

$$\int_{\mathbb{R}^q} \varphi(u) T(u, v)\, dv\, du.$$

The existence of these integrals, as well as their equality, is guaranteed by the First Fubini Theorem.

Appendix 1

Partitions of unity

Partitions of unity provide a method for decomposing a distribution T defined on a 'large' compact set K into a sum of distributions T_i, each of which is supported on 'small' neighborhood U_i within K.

We recall that, by the definition of compactness, any covering of a compact set K by arbitrary open neighborhoods U_i can be replaced by a finite subcovering. That is, there exists a finite collection of the open sets U_i whose union contains (or 'covers') K.

Let $\Theta = [U_1, \ldots, U_n]$ be a finite open covering of a compact set K in $\mathbb{R}^q$. A *partition of unity* for the pair $\langle \Theta, K \rangle$ is a sequence $\psi_1, \ldots, \psi_n$ of C^∞ functions such that

(i) $0 \leqslant \psi_i \leqslant 1$;

(ii) $\psi_i(x) \equiv 0$ for $x \notin U_i$;

(iii) $\sum_{i=1}^n \psi_i(x) \equiv 1$ for $x \in K$.

Theorem. A partition of unity exists for every $\langle \Theta, K \rangle$.

Corollary. A distribution T with support on K can be decomposed:

$$T = \sum_{i=1}^n T_i,$$

where each T_i has support on U_i.

(Proof of corollary: take $T_i = \psi_i T$, with ψ_i as above.)

Proof of theorem. For clarity of presentation, we will give the proof for $\mathbb{R}^1$. The extension to $\mathbb{R}^q$ can be done merely by considering q-cubes in place of intervals. Take a $\delta > 0$ such that every interval of length 2δ which intersects K is contained in some U_i. Choose a and N so that the interval $(a + \delta, a + (N - 1)\delta) \supset K$.

Let $I_i \equiv (a + (i-1)\delta, a + (i+1)\delta)$, $0 \leqslant i \leqslant N$. For each I_i, choose a function $\varphi_i \in C^\infty$ such that $\varphi_i(x) > 0$ for $x \in I_i$, $\varphi_i(x) = 0$ for $x \notin I_i$.

Let J be the set of integers i such that $I_i \cap K \neq \varnothing$ (this implies $1 \leqslant i \leqslant N-1$). For $i \in J$, let

$$\psi_i(x) \equiv \varphi_i(x) \bigg/ \sum_{j=0}^{N} \varphi_j(x).$$

Then $\psi_i \in C^\infty$ (since $i \neq 0, N$), and $\sum_{i \in J} \psi_i(x) \equiv 1$ for $x \in K$. But each ψ_i has support on one of the sets U_j.   Q.E.D.

Appendix 2

The Structure Theorem

Here we prove a basic structure theorem for tempered distributions. This theorem asserts that every tempered distribution is a finite derivative of a slowly increasing continuous function.

Remarks. Of course, the continuous function need not be differentiable in the classical sense. Its derivatives exist only in the sense of distribution theory, and they are not, in general, functions. This is nothing more than the assertion – abundantly clear by now – that distributions are more general than functions. The Structure Theorem asserts that, in some sense, distributions are not *too* general. They all come from ordinary continuous functions, via the distribution differentiation process.

(There is a similar, but slightly more complicated, structure theorem for general distributions. We shall not give it here. For this theorem, cf. e.g., [ED].)

To prove the Structure Theorem, we will need to use a little more measure theory than previously. Specifically, we need the well known Riesz Representation Theorem, which asserts that every bounded linear functional on $C[a, b]$ corresponds to a bounded complex measure μ on $[a, b]$.

(A *bounded complex measure* is a measure μ of the form $\mu = \mu_1 - \mu_2 + i\mu_3 - i\mu_4$, where the μ_i are finite positive measures. A *bounded linear functional* on $C[a, b]$ is a linear mapping L from $C[a, b]$ into the complex numbers which satisfies $|L(f)| \leqslant \text{Const} \cdot \|f\|_\infty$ for some universal Const and all $f \in C[a, b]$; of course $\|f\|_\infty$ means $\max\{|f(x)|: x \in [a, b]\}$.)

We now state the Riesz Theorem without proof. (For proofs, see e.g., [HA] and [LO].)

Riesz Representation Theorem. Every bounded linear functional L on $C[a, b]$ is related to a bounded complex measure μ on $[a, b]$ via the equation

$$L(f) = \int_a^b f(x)\, d\mu(x), \quad \text{for all } f \in C[a, b].$$

We will also need the Hahn–Banach Theorem, which asserts that any bounded linear functional defined on a subspace Y of a Banach space X can be extended to a bounded linear functional defined on X.

Recall that a function $g(x)$ on $\mathbb{R}^q$ is *slowly increasing* if there exists an integer N such that $|g(x)| \leqslant \text{Const}(1 + |x|)^N$. Similarly, a complex measure $d\mu(x)$ is *slowly increasing* if $\int_{\mathbb{R}^q} (1 + |x|)^{-N}|d\mu(x)| < \infty$ for some N.

Our theorems are:

Structure Theorem – first form. Every distribution $T \in \mathscr{S}'(\mathbb{R}^q)$ is a finite mixed partial derivative of a slowly increasing continuous function.

The following variation is sometimes useful.

Structure Theorem – second form. For every tempered distribution T on $\mathbb{R}^q$, there exists an integer M, and slowly increasing complex measures μ_α on $\mathbb{R}^q$, corresponding to the finite set of multi indices α with $|\alpha| \leqslant M$, such that

$$T(x) = \sum_{|\alpha| \leqslant M} D^\alpha \mu_\alpha(x).$$

Before proceeding to the main proofs, we give a lemma which encompasses the continuity condition in the definition of a tempered distribution.

Lemma. Let T be a tempered distribution. Then there exist integers M and N and a constant $A > 0$ such that for all test functions φ,

$$|\langle T, \varphi \rangle| \leqslant A \cdot \sum_{|\alpha| \leqslant M} \|(1 + |x|)^N \varphi^{(\alpha)}(x)\|_\infty.$$

Proof. Suppose this is false. Then, for every M, N and A, there exists a test function φ with $\|(1 + |x|)^N \varphi^{(\alpha)}(x)\|_\infty < 1/A$ $(0 \leqslant |\alpha| \leqslant M)$, but $\langle T, \varphi \rangle = 1$. For $n = 1, 2, 3, \ldots$, let φ_n be a function which satisfies these conditions with $M = N = A = n$. Then the sequence $\{\varphi_n\}$ converges to zero in the test function space $\mathscr{S}(\mathbb{R}^q)$, but $\langle T, \varphi_n \rangle = 1$ for all n. This contradicts the continuity of T.

                                                                    Q.E.D.

For clarity we will give the remainder of the proof for $\mathbb{R}^1$. The extension to $\mathbb{R}^q$ is trivial, involving only a proliferation of subscripts. On $\mathbb{R}^1$, the finite set of measures μ_α is replaced by the sequence $\mu_0, \ldots, \mu_M$, corresponding to $(d/dx)^k$, $0 \leqslant k \leqslant M$.

Proof of the Structure Theorem – second form. We restate the theorem slightly.

Suppose a distribution T satisfies the conclusion of lemma, for a given M and N, and for all test functions ϕ with *compact support* (i.e. for $\phi \in \mathcal{D}$). Then there exist complex measures $\mu_0, \ldots, \mu_M$ such that

$$T = \sum_{k=0}^{M} (d/dx)^k \mu_k,$$

and

$$\int_{-\infty}^{\infty} \frac{|d\mu_k(x)|}{(1 + |x|)^N} < \infty, \qquad 0 \leqslant k \leqslant M.$$

We now prove this modified theorem. Let S be the *union* (not a cartesian product) of $M + 1$ disjoint copies $R_0, \ldots, R_M$ of the real line. (S could be viewed as a subset in $\mathbb{R}^2$ consisting of the $M + 1$ lines $y = 0$, $y = 1, \ldots, y = M$ in the conventional (x, y) coordinate system.) Let X be the Banach space of all continuous complex valued functions $\phi = (\phi_0, \ldots, \phi_M)$ on S ($\phi_i = \phi | R_i$) for which the norm,

$$\||\phi\|| \equiv \sup_k \|(1 + |x|)^N \phi_k(x)\|_\infty,$$

is finite. Let Y be the (non-closed) subspace consisting of all $(M + 1)$-tuples $(\phi, \phi', \phi'', \ldots, \phi^{(M)})$ (i.e. with $\phi_{i+1} = d\phi_i/dx$), where $\phi \in \mathcal{D}(\mathbb{R}^1)$.

(Note that, because of the constraint, the closure $\bar{Y}$ is a proper subspace of X.)

Then, by our assumptions, the distribution T defines a bounded linear functional on Y, via $T(\phi, \phi', \ldots, \phi^M) = \langle T, \phi \rangle$, and by the *Hahn–Banach Theorem*, T has an extension to a bounded linear functional T^* on X.

Now let T_0^* be the restriction of T^* to $C_o(S)$, continuous functions on S which vanish at infinity (not $\mathcal{D}$). The *Riesz Representation Theorem* implies that there is a *locally finite* complex measure μ on S such that $T_0^*(\phi) = \int_S \phi \, d\mu$ for all $\phi \in C_o(S)$.

Define $\mu_k \equiv \mu | R_k$, $0 \leqslant k \leqslant M$.

Let $v_k(x) \equiv \mu_k(x)/(1 + |x|)^N$.

Then v_k is a complex measure on the line such that

$$\int_{-\infty}^{\infty} \phi \, dv_k \leqslant A \|\phi\|_\infty \text{ for all } \phi \in C_o(R^1).$$

(Proof: T_0^* is bounded in terms of $\||\ \||$; hence just apply T_0^* to a function ϕ on S such that $\phi_i(x) = \phi(x)/(1 + |x|)^N$ if $i = k$, $\phi_i(x) = 0$ for $i \neq k$).

But this implies

$$\int_{-\infty}^{\infty} |dv_k(x)| \leqslant A,$$

whence

$$\int_{-\infty}^{\infty} |d\mu_k(x)|/(1 + |x|)^N \leqslant A. \tag{*}$$

Now, by definition of μ, for all test function ϕ with *compact support*,

$$\langle T, \phi \rangle = \sum_{k=0}^{M} \int_{-\infty}^{\infty} \phi^{(k)}(x) \, d\mu_k(x). \tag{**}$$

But $\mathscr{D}$ is *dense* in $\mathscr{S}$, and T is continuous in terms of the topology on $\mathscr{S}$. Now the *inequality* (*) *shows that the functional on the right hand side of* (**) *is also continuous on* $\mathscr{S}$. Hence the identity (**) holds for all $\phi \in \mathscr{S}$. The Structure Theorem – second form is proved.

(This involves a slight change of notation; here

$$T = \sum_{k=0}^{M} (-1)^k (d/dx)^k \mu_k.)$$

Proof of Structure Theorem – second form. The Structure Theorem – second form gives the first form upon integrating each measure μ_k, $0 \leqslant k \leqslant M$, $(M + 2 - k)$ times to get a continuous function g_k, and setting $g = \sum g_k$; then

$$T = \sum_{k=0}^{M} \mu_k^{(k)} = (d/dx)^{M+2} g.$$

Appendix 3

Proof of Theorems A and B from Chapter 7

Here we prove the two results, Theorems A and B, which were left unproved in Chapter 7, Section 3. The reason for postponing these proofs is that they are much easier to do once we have the Structure Theorem of Appendix 2.

We follow the notations set in Chapter 7. Thus U and V are complementary subspaces of $\mathbb{R}^q$ with dimensions r and s, $r + s = q$, respectively.

Lemma. Let T_f and T_g be distributions corresponding to slowly increasing continuous functions $f(u)$ on U and $g(v)$ on V, respectively. Let $\alpha = (\alpha_1, \ldots, \alpha_r)$ and $\beta = (\beta_1, \ldots, \beta_s)$ be multi-indices. Then the tensor product $T_f^{(\alpha)} T_g^{(\beta)}$ is a distribution in $\mathscr{S}'(\mathbb{R}^q)$ and for any $\varphi(u, v) \in \mathscr{S}(\mathbb{R}^q)$

$$\langle T_f^{(\alpha)} T_g^{(\beta)}, \varphi(u, v) \rangle = (-1)^{|\alpha| + |\beta|} \langle T_f, \langle T_g, \varphi^{(\alpha, \beta)}(u, v) \rangle \rangle$$

$$= (-1)^{|\alpha| + |\beta|} \int_{\mathbb{R}^q} f(u) g(v) \varphi^{(\alpha, \beta)}(u, v) \, dv \, du$$

$$= \langle T_{f(u)g(v)}^{(\alpha, \beta)}, \varphi(u, v) \rangle$$

Proof. Let $\psi(u) = \langle T_g^{(\beta)}(v), \varphi(u, v) \rangle$ and let us first show that

$$\psi(u) \in \mathscr{S}(U) \quad \text{and} \quad D^\alpha \psi(u) = (-1)^{|\beta|} \langle T_g, \varphi^{(\alpha, \beta)}(u, v) \rangle. \qquad (*)$$

By the definition of distribution derivative, $\psi(u) = (-1)^{|\beta|} \langle T_g(v), \varphi^{(0, \beta)}(u, v) \rangle$ and, since g is a slowly increasing continuous function on V,

$$\psi(u) = (-1)^{|\beta|} \int_V g(v) \varphi^{(0, \beta)}(u, v) \, dv.$$

Now, again, since g is a slowly increasing continuous function, and since $\varphi^{(0, \beta)}$ is a rapidly decreasing C^∞ function, the integral is absolutely convergent. By a well-known theorem in advanced calculus, we can take any partial derivative of $\psi(u)$ by interchanging the derivative which is in the u-variable with the integral which is in the v-variable. Hence $\psi(u)$ is C^∞. In particular,

$$\psi^{(\alpha)}(u) = (-1)^{|\beta|} \int_V g(v)\varphi^{(\alpha,\beta)}(u, v)\, dv = (-1)^{|\beta|} \langle T_g, \varphi^{(\alpha,\beta)} \rangle.$$

The boundedness of $|u|^N \psi^\gamma(u)$ for any integer N and any multi-index γ follows by bringing $|u|^N$ inside the integral and noting that $|u|^N \varphi^{(\alpha,\beta)}(u, v)$ is rapidly decreasing. Hence, $\psi(u) \in \mathscr{S}(U)$ and $\psi^{(\alpha)}(u) = (-1)^{|\beta|} \langle T_g(v), \varphi^{(\alpha,\beta)}(u, v) \rangle$. This proves (*).

Now we continue,

$$\begin{aligned}
\langle T_f^{(\alpha)}(u), \langle T_g^{(\beta)}(v), \varphi(u, v) \rangle \rangle &= \langle T_f^{(\alpha)}(u), \psi(u) \rangle \\
&= (-1)^{|\alpha|} \langle T_f(u), \psi^{(\alpha)}(u) \rangle \\
&= (-1)^{|\alpha|+|\beta|} \langle T_f, \langle T_g, \varphi^{(\alpha,\beta)} \rangle \rangle \\
&= (-1)^{|\alpha|+|\beta|} \int_U f(u) \int_V g(v)\varphi^{(\alpha,\beta)}(u, v)\, dv\, dv \\
&= (-1)^{|\alpha|+|\beta|} \int_{\mathbb{R}^q} f(u)g(v)\varphi^{(\alpha,\beta)}(u, v)\, dv\, du.
\end{aligned}$$

The last equality follows from Fubini's Theorem, since $f(u)g(v)$ is slowly increasing and $\varphi^{(\alpha,\beta)}(u, v)$ is rapidly decreasing.

(The Fubini Theorem used here is, of course, the classical Fubini Theorem for multiple integrals. It is not the more abstract 'Fubini Theorem' proved in Chapter 7.)

Since $f(u)g(v)$ is a slowly increasing continuous function on $\mathbb{R}^q$, it defines a distribution T_{fg} in $\mathbb{R}^q$ and

$$\begin{aligned}
\langle T_{fg}^{(\alpha,\beta)}, \varphi \rangle &= (-1)^{|\alpha|+|\beta|} \langle T_{fg}, \varphi^{(\alpha,\beta)} \rangle \\
&= (-1)^{|\alpha|+|\beta|} \int_{\mathbb{R}^q} f(u)g(v)\varphi^{(\alpha,\beta)}(u, v)\, dv\, du.
\end{aligned}$$

The previous chain of equations gave us

$$\langle T_f^{(\alpha)} T_g^{(\beta)}, \varphi(u, v) \rangle = (-1)^{|\alpha|+|\beta|} \int_{\mathbb{R}^q} f(u)g(v)\varphi^{(\alpha,\beta)}(u, v)\, dv\, du.$$

Comparing this with the above, we obtain

$$T_f^{(\alpha)} T_g^{(\beta)} = T_{fg}^{(\alpha,\beta)},$$

together with all of the desired intermediate equalities. This proves the lemma.

Theorem A. The tensor product $S(u)T(v)$ is a tempered distribution on $\mathbb{R}^q$.

Theorem B. The tensor product is commutative, i.e. $S(u)T(v) = T(v)S(u)$.

The above theorems are corollaries of the lemma and the Structure Theorem – first form.

Proof of Theorem A. By the Structure Theorem – first form, there are slowly increasing continuous functions $f(u)$ and $g(v)$ such that $S(u) = T_f^{(\alpha)}(u)$ and $T(v) = T_g^{(\beta)}(v)$ for some multi-indices α and β. Hence the theorem follows from the above lemma.

Proof of Theorem B. Following the Structure Theorem – first form, let $S(u) = T_f^{(\alpha)}$ and $T(v) = T_g^{(\beta)}$ for some slowly increasing continuous functions $f(u)$ and $g(v)$. Then, by the lemma, for any $\varphi \in \mathscr{S}(\mathbb{R}^q)$,

$$\langle S(u)T(v), \varphi(u, v) \rangle = (-1)^{|\alpha| + |\beta|} \int_{\mathbb{R}^q} f(u)g(v)\varphi^{(\alpha, \beta)}(u, v) \, dv \, du$$

$$= (-1)^{|\alpha| + |\beta|} \int_{\mathbb{R}^q} g(v)f(u)\varphi^{(\alpha, \beta)}(u, v) \, dv \, du$$

$$= \langle T(v)S(u), \varphi(u, v) \rangle.$$

BIBLIOGRAPHY

Those references cited in the text are preceded by the appropriate code used in the text.

Classical analysis

For the purposes of this bibliography, books not involving distribution theory are considered 'classical'.

Ahlfors, L. V. (1953), *Complex Analysis*, McGraw-Hill, New York.

Buck, R. C. (with the collaboration of E. F. Buck) (1978), *Advanced Calculus*, 3rd edn., McGraw-Hill, New York.

Churchill, R. V. (1963), *Fourier Series and Boundary Value Problems*, 2nd edn., McGraw-Hill, New York.

Churchill, R. V. (1972), *Operational Mathematics*, 3rd edn., McGraw-Hill, New York.

Coddington, E. A. and Levinson, N. (1955), *Theory of Ordinary Differential Equations*, McGraw-Hill, New York.

[DUS]: Dunford, N. and Schwartz, J. T. (1958), *Linear Operators*, Part I, Interscience, New York.

[HA]: Halmos, P. R. (1950), *Measure Theory*, D. Van Nostrand, New York.

Halmos, P. R. (1958), *Finite-Dimensional Vector Spaces*, 2nd edn., D. Van Nostrand, Princeton.

[HR]: Herstein, I. N. (1964), *Topics in Algebra*, Blaisdell, New York.

[LO]: Loomis, L. H. (1953), *An Introduction to Abstract Harmonic Analysis*, D. Van Nostrand, New York.

Paley, R. E. A. C. and Wiener, N. (1934), *Fourier Transforms in the Complex Domain*, American Mathematical Society, New York.

Rudin, W. (1976), *Principles of Mathematical Analysis*, 3rd edn., McGraw-Hill, New York.

Rudin, W. (1987), *Real and Complex Analysis*, 3rd edn., McGraw-Hill, New York.

[WHW]: Whittaker, E. T. and Watson, G. N. (1963), *A Course of Modern Analysis*, 4th edn., Cambridge University Press.

Widder, D. V. (1941), *The Laplace Transform*, Princeton University Press.

Wiener, N. (1933), *The Fourier Integral and Certain of its Applications*, Cambridge University Press.

Distribution theory

The following books either deal exclusively with distributions or make heavy use of them.

Bremermann, H. (1965), *Distributions, Complex Variables, and Fourier Transforms*, Addison-Wesley, Reading, Mass.

Chevalley, C. (1950), *Theory of Distributions*, Lectures at Columbia University, Columbia University Press.
[ED]: Edwards, R. E. (1965), *Functional Analysis*, Holt, Rinehart and Winston, New York.
Friedlander, F. G. (1982), *Introduction to the Theory of Distributions*, Cambridge University Press.
Friedman, A. (1963), *Generalized Functions and Partial Differential Equations*, Prentice-Hall, Englewood Cliffs, N.J.
[GSV]: Gelfand, I. M. and Šilov, G. (1964–), *Generalized Functions*, 4 vols., Academic Press, New York.
[HO]: Hörmander, L. (1964), *Linear Partial Differential Operators*, Springer-Verlag, Academic Press Inc. Publishers, New York.
Horváth, J. (1966), *Topological Vector Spaces and Distributions*, Addison-Wesley, Reading, Mass.
Katznelson, Y. (1968), *An Introduction to Harmonic Analysis*, John Wiley and Sons, New York.
Kolmogorov, A. N. and Fomin, S. V. (1957), *Elements of the Theory of Functions and Functional Analysis* (translated by L. F. Boron), Graylock Press, Rochester, N.Y.
[LI]: Lighthill, M. J. (1958), *Introduction to Fourier Analysis and Generalized Functions*, Cambridge University Press.
Rudin, W. (1973), *Functional Analysis*, McGraw-Hill, New York.
[SW]: Schwartz, L. (1950, 1951), *Théorie des Distributions*, vols. I et II, Hermann et Cie, Paris.
Treves, F. (1967), *Topological Vector Spaces, Distributions and Kernels*, Academic Press, New York.
Yosida, K. (1968), *Functional Analysis*, 2nd edn., Springer-Verlag, New York.
Zemanian, A. H. (1965), *Distribution Theory and Transform Analysis; An Introduction to Generalized Functions with Applications*, McGraw-Hill, New York.

Research articles

The research literature on distribution theory is vast. We have simply listed a few papers which touch on Chapter 7 – the only chapter in this book which approaches the research level.

[HV]: Horváth, J. (1974), 'Sur la convolution des distributions', *Bull. Soc. Math.*, 2ᵉ Série, 98, 183–92.
[KO]: König, H. (1955), 'Multiplikation von Distributionen', *Math. Ann.* 128, 420–52.
Schwartz, L. (1953–4), 'Produits tensoriels topologiques et espaces nucléaires', Séminaire, Institut Henri Poincarè, Paris.
[SR]: Shiraishi, R. (1959), 'On the definition of convolution for distributions', *J. Sci. Hiroshima Univ.*, Ser. A-23, 19–32.
Youn, H. K. and Richards, I. (1980), 'On the general definition of convolution for distributions', *J. Korean Math. Soc.* 17, no. 1, 13–37.
Youn, H. K. and Richards, I. (1981), 'On the definition of convolution for several distributions', *J. Korean Math. Soc.* 17, no. 2, 161–8.

INDEX

Printed in the United States
By Bookmasters